T0155656

# Reinforcement Learning for Finance

## Solve Problems in Finance with CNN and RNN Using the TensorFlow Library

Samit Ahlawat

Apress®

*Reinforcement Learning for Finance: Solve Problems in Finance with CNN and RNN Using the TensorFlow Library*

Samit Ahlawat
Irvington, NJ, USA

ISBN-13 (pbk): 978-1-4842-8834-4
https://doi.org/10.1007/978-1-4842-8835-1

ISBN-13 (electronic): 978-1-4842-8835-1

Managing Director, Apress Media LLC: Welmoed Spahr
Acquisitions Editor: Celestin Suresh John
Development Editor: Laura Berendson
Coordinating Editor: Mark Powers

Cover designed by eStudioCalamar

Cover image by Joel Filipe on Unsplash (www.unsplash.com)

Distributed to the book trade worldwide by Apress Media, LLC, 1 New York Plaza, New York, NY 10004, U.S.A. Phone 1-800-SPRINGER, fax (201) 348-4505, e-mail orders-ny@springer-sbm.com, or visit www.springeronline.com. Apress Media, LLC is a California LLC and the sole member (owner) is Springer Science + Business Media Finance Inc (SSBM Finance Inc). SSBM Finance Inc is a **Delaware** corporation.

For information on translations, please e-mail booktranslations@springernature.com; for reprint, paperback, or audio rights, please e-mail bookpermissions@springernature.com.

Apress titles may be purchased in bulk for academic, corporate, or promotional use. eBook versions and licenses are also available for most titles. For more information, reference our Print and eBook Bulk Sales web page at http://www.apress.com/bulk-sales.

Any source code or other supplementary material referenced by the author in this book is available to readers on GitHub (https://github.com/Apress). For more detailed information, please visit http://www.apress.com/source-code.

Printed on acid-free paper

*To my family and friends without whose support this book would not have been possible.*

# Table of Contents

TABLE OF CONTENTS

# About the Author

 **Samit Ahlawat** is Senior Vice President in Quantitative Research, Capital Modeling, at JPMorgan Chase in New York, USA. In his current role, he is responsible for building trading strategies for asset management and for building risk management models. His research interests include artificial intelligence, risk management, and algorithmic trading strategies. He has given CQF Institute talks on artificial intelligence, has authored several research papers in finance, and holds a patent for facial recognition technology. In his spare time, he contributes to open source code.

# Acknowledgments

I would like to express my heartfelt appreciation for my friends and coworkers, in academia and the workplace, who encouraged me to write this book.

# Preface

When I began using artificial intelligence tools in quantitative financial research, I could not find a comprehensive introductory text focusing on financial applications. Neural network libraries like TensorFlow, PyTorch, and Caffe had made tremendous contributions in the rapid development, testing, and deployment of deep neural networks, but I found most applications restricted to computer science, computer vision, and robotics. Having to use reinforcement learning algorithms in finance served as another reminder of the paucity of texts in this field. Furthermore, I found myself referring to scholarly articles and papers for mathematical proofs of new reinforcement learning algorithms. This led me to write this book to provide a one-stop resource for Python programmers to learn the theory behind reinforcement learning, augmented with practical examples drawn from the field of finance.

In practical applications, reinforcement learning draws upon deep neural networks. To facilitate exposition of topics in reinforcement learning and for continuity, this book also provides an introduction to TensorFlow and covers neural network topics like convolutional neural networks (CNNs) and recurrent neural networks (RNNs).

Finally, this book also introduces readers to writing modular, reusable, and extensible reinforcement learning code. Having worked on developing trading strategies using reinforcement learning and publishing papers, I felt existing reinforcement learning libraries like TF-Agents are tightly coupled with the underlying implementation framework and do not express central concepts in reinforcement learning in a manner that is modular enough for someone conversant with concepts to pick up TF-Agent library usage or extend its algorithms for specific applications. The code samples covered in this book provide examples of how to write modular code for reinforcement learning.

# Introduction

Reinforcement learning is a rapidly growing area of artificial intelligence that involves an agent learning from past experience of rewards gained by taking specific actions in certain states. The agent seeks to learn a policy prescribing the optimum action in each state with the objective of maximizing expected discounted future rewards. It is an unsupervised learning technique where the agent learns the optimum policy by past interactions with the environment. Supervised learning, by contrast, seeks to learn the pattern of output corresponding to each state in training data. It attempts to train the model parameters in order to get a close correspondence between predicted and actual output for a given set of inputs. This book outlines the theory behind reinforcement learning and illustrates it with examples of implementations using TensorFlow. The examples demonstrate the theory and implementation details of the algorithms, supplemented with a discussion of corresponding APIs from TensorFlow and examples drawn from quantitative finance. It guides a reader familiar with Python programming from basic to advanced understanding of reinforcement learning algorithms, coupled with a comprehensive discussion on how to use state-of-the-art software libraries to implement advanced algorithms in reinforcement learning.

Most applications of reinforcement learning have focused on robotics or computer science tasks. By focusing on examples drawn from finance, this book illustrates a spectrum of financial applications that can benefit from reinforcement learning.

# CHAPTER 1

# Overview

Deep neural networks have transformed virtually every scientific human endeavor – from image recognition, medical imaging, robotics, and self-driving cars to space exploration. The extent of transformation heralded by neural networks is unrivaled in contemporary human history, judging by the range of new products that leverage neural networks. Smartphones, smartwatches, and digital assistants – to name a few – demonstrate the promise of neural networks and signal their emergence as a mainstream technology. The rapid development of artificial intelligence and machine learning algorithms has coincided with increasing computational power, enabling them to run rapidly. Keeping pace with new developments in this field, various open source libraries implementing neural networks have blossomed. Python has emerged as the *lingua franca* of the artificial intelligence programming community. This book aims to equip Python-proficient programmers with a comprehensive knowledge on how to use the TensorFlow library for coding deep neural networks and reinforcement learning algorithms effectively. It achieves this by providing detailed mathematical proofs of key theorems, supplemented by implementation of those algorithms to solve real-life problems.

Finance has been an early adopter of artificial intelligence algorithms with the application of neural networks in designing trading strategies as early as the 1980s. For example, White (1988) applied a simple neural network to find nonlinear patterns in IBM stock price. However, recent cutting-edge research on reinforcement learning has focused

© Samit Ahlawat 2023
S. Ahlawat, *Reinforcement Learning for Finance*,
https://doi.org/10.1007/978-1-4842-8835-1_1

predominantly on robotics, computer science, or interactive game-playing. The lack of financial applications has led many to question the applicability of deep neural networks in finance where traditional quantitative models are ubiquitous. Finance practitioners feel that the lack of rigorous mathematical proofs and transparency about how neural networks work has restricted their wider adoption within finance. This book aims to address both of these concerns by focusing on real-life financial applications of neural networks.

# 1.1 Methods for Training Neural Networks

Neural networks can be trained using one of the following three methods:

1. **Supervised learning** involves using a training dataset with known output, also called ground truth values. For a classification task, this would be the true labels, while for a regression task, it would be the actual output value. A loss function is formulated that measures the deviation of the model output from the true output. This function is minimized with respect to model parameters using stochastic gradient descent.

2. **Unsupervised learning** methods use a training dataset made up of input features without any knowledge of the true output values. The objective is to classify inputs into clusters for clustering or dimension reduction applications or for identifying outliers.

3. **Reinforcement learning** involves an agent that learns an optimal policy within the framework of a Markov decision problem (MDP). The training

dataset consists of a set of actions taken in different states by an agent, followed by rewards earned and the next state to which the agent transitions. Using the history of rewards, reinforcement learning attempts to learn an optimal policy to maximize the expected sum of discounted future rewards. This book focuses on reinforcement learning.

# 1.2  Machine Learning in Finance

Machine learning applications in finance date back to the 1980s with the use of neural networks in stock price prediction (White, 1988). Within finance, automated trading strategies and portfolio management have been early adopters of artificial intelligence and machine learning tools. Allen and Karjalainen (1999) applied genetic algorithms to combine simple trading rules to form more complex ones. More recent applications of machine learning in finance can be seen in the works of Savin et al. (2007), who used the pattern recognition method presented by Lo et al. (2000) to test if the head-and-shoulders pattern had predictive power; Chavarnakul and Enke (2008), who employed a generalized regression neural network (GRNN) to construct two trading strategies based on equivolume charting that predicted the next day's price using volume- and price-based technical indicators; and Ahlawat (2016), who applied probabilistic neural networks to predict technical patterns in stock prices. Other works include Enke and Thawornwong (2005), Li and Kuo (2008), and Leigh et al. (2005). Chenoweth et al. (1996) have studied the application of neural networks in finance. Enke and Thawornwong (2005) tested the hypothesis that neural networks can provide superior prediction of future returns based on their ability to identify nonlinear relationships. They employed only fundamental measures and did not consider technical ones. Their neural network provided higher returns than the buy-and-hold strategy, but they did not consider transaction costs.

There are many other applications of machine learning in finance besides trading strategies, perhaps less glamorous but equally significant in business impact. This book gives a comprehensive exposition of several machine learning applications in finance that are at cutting edge of research and practical use.

## 1.3  Structure of the Book

This book begins with an introduction to the TensorFlow library in Chapter 2 and illustrates the concepts with financial applications that involve building models to solve practical problems. The datasets for problems are publicly available. Relevant concepts are illustrated with mathematical equations and concise explanations.

Chapter 3 introduces readers to convolutional neural networks (CNNs), and Chapter 4 follows up with a similar treatment of recurrent neural networks (RNNs). These networks are frequently used in building value function models and policies in reinforcement learning, and a comprehensive understanding of CNN and RNN is indispensable for using reinforcement learning effectively on practical problems. As before, all foundational concepts are illustrated with mathematical theory, explanation, and practical implementation examples.

Chapter 5 introduces reinforcement learning concepts: from Markov decision problem (MDP) formulation to defining value function and policies, followed by a comprehensive discussion of reinforcement learning algorithms illustrated with examples and mathematical proofs.

Finally, Chapter 6 provides a discussion of recent, groundbreaking advances in reinforcement learning by discussing technical papers and applying those algorithms to practical applications.

# CHAPTER 2

# Introduction to TensorFlow

TensorFlow is an open source, high-performance machine learning library developed by Google and released for public use in 2015. It has interfaces for Python, C++, and Java programming languages. It has the option of running on multiple CPUs or GPUs. TensorFlow offers two modes of execution: eager mode that can be run immediately and graph mode that creates a dependency graph and executes nodes in that graph only where needed.

This book uses TensorFlow 2.9.1. Older TensorFlow constructs from version 1 of the library such as **Session** and **placeholder** are not covered here. Their use has been rendered obsolete in TensorFlow version 2.0 and higher. Output shown in the code listings has been generated using the PyCharm IDE's interactive shell.

## 2.1 Tensors and Variables

Tensors are n-dimensional arrays, similar in functionality to the numpy library's ndarray object. They are instances of the **tf.Tensor** object. A three-dimensional tensor of 32-bit floating-point numbers can be created using code in Listing 2-1. Tensor has attributes shape and dtype that tell the shape and data type of the tensor. Once created, tensors retain their shape.

© Samit Ahlawat 2023
S. Ahlawat, *Reinforcement Learning for Finance*,
https://doi.org/10.1007/978-1-4842-8835-1_2

***Listing 2-1.*** Creating a Three-Dimensional Tensor

```
1    import tensorflow as tf
2
3    tensor = tf.constant([[list(range(3))],
4                          [list(range(1, 4))],
5                          [list(range(2, 5))]], dtype=tf.
                           float32)
6
7    print(tensor)
8
9    tf.Tensor(
10   [[[0. 1. 2.]]
11   [[1. 2. 3.]]
12   [[2. 3. 4.]]], shape=(3, 1, 3), dtype=float32)
```

Most numpy functions for creating ndarrays have analogs in TensorFlow, for example, tf.ones, tf.zeros, tf.eye, tf.ones_like, etc. Tensors support usual mathematical operations like +, −, etc., in addition to matrix operations like transpose, matmul, and einsum, as shown in Listing 2-2.

***Listing 2-2.*** Mathematical Operations on Tensors

```
1    import tensorflow as tf
2
3    ar = tf.constant([[1, 2], [2, 2]], dtype=tf.float32)
4
5    print(ar)
6    <tf.Tensor: id=1, shape=(2, 2), dtype=float32, numpy=
7    array([[1., 2.],
8    [2., 2.]], dtype=float32)>
9
10   # elementwise multiplication
```

```
11   print(ar * ar)
12   Out[8]:
13   <tf.Tensor: id=2, shape=(2, 2), dtype=float32, numpy=
14   array([[1., 4.],
15   [4., 4.]], dtype=float32)>
16
17   # matrix multiplication C = tf.matmul(A, B) => cij =
     sum_k (aik * bkj)
18   print(tf.matmul(ar, tf.transpose(ar)))
19
20   <tf.Tensor: id=5, shape=(2, 2), dtype=float32, numpy=
21   array([[5., 6.],
22   [6., 8.]], dtype=float32)>
23
24   # generic way of matrix multiplication
25   print(tf.einsum("ij,kj->ik", ar, ar))
26
27   <tf.Tensor: id=23, shape=(2, 2), dtype=float32, numpy=
28   array([[5., 6.],
29   [6., 8.]], dtype=float32)>
30
31   # cross product
32   print(tf.einsum("ij,kl->ijkl", ar, ar))
33
34   <tf.Tensor: id=32, shape=(2, 2, 2, 2),
     dtype=float32, numpy=
35   array([[[[1., 2.],
36   [2., 2.]],
37   [[2., 4.],
38   [4., 4.]]],
39   [[[2., 4.],
```

```
40    [4., 4.]],
41    [[2., 4.],
42    [4., 4.]]]], dtype=float32)>
```

Tensors can be sliced using the usual Python notation with a semicolon. For advanced slicing, use **tf.slice** that accepts a begin index and the number of elements along each axis to slice. **tf.strided_slice** can be used for adding a stride. To obtain specific indices from a tensor, use **tf.gather**. To extract specific elements of a multidimensional tensor specified by a list of indices, use **tf.gather_nd**. These APIs are illustrated using examples in Listing 2-3.

***Listing 2-3.*** Tensor Slicing Operations

```
1    import tensorflow as tf
2
3    tensor = tf.constant([[1, 2], [2, 2]], dtype=tf.float32)
4
5    print(tensor[1:, :])
6    <tf.Tensor: id=37, shape=(1, 2), dtype=float32,
     numpy=array([[2., 2.]], dtype=float32)>
7
8    print(tf.slice(tensor, begin=[0,1], size=[2, 1]))
9    tf.Tensor(
10   [[2.]
11   [2.]], shape=(2, 1), dtype=float32)
12
13   print(tf.gather_nd(tensor, indices=[[0, 1], [1, 0]]))
14   Out[18]: <tf.Tensor: id=42, shape=(2,), dtype=float32,
     numpy=array([2., 2.], dtype=float32)>
```

Ragged tensors are tensors with a nonuniform shape along an axis, as illustrated in Listing 2-4.

**Listing 2-4.** Ragged Tensors

```
1    import tensorflow as tf
2
3    jagged = tf.ragged.constant([[1, 2], [2]])
4    print(jagged)
5    <tf.RaggedTensor [[1, 2], [2]]>
```

TensorFlow allows space-efficient storage of sparse arrays, that is, arrays with most elements as 0. The **tf.sparse.SparseTensor** API takes the indices of non-zero elements, their values, and the dense shape of the sparse array. This is shown in Listing 2-5.

**Listing 2-5.** Sparse Tensors

```
1    import tensorflow as tf
2
3    tensor = tf.sparse.SparseTensor(indices=[[1,0], [2,2]],
     values=[1, 2], dense_shape=[3, 4])
4    print(tensor)
5    SparseTensor(indices=tf.Tensor(
6    [[1 0]
7    [2 2]], shape=(2, 2), dtype=int64), values=tf.Tensor([1 2],
     shape=(2,), dtype=int32), dense_shape=tf.Tensor([3 4],
     shape=(2,), dtype=int64))
8
9    print(tf.sparse.to_dense(tensor))
10   tf.Tensor(
11   [[0 0 0 0]
12   [1 0 0 0]
13   [0 0 2 0]], shape=(3, 4), dtype=int32)
```

In contrast to **tf.Tensor** that is immutable after creation, a TensorFlow variable can be changed. A variable is an instance of the **tf.Variable** class and can be created by initializing it with a tensor. Variables can be converted to tensors using **tf.convert_to_tensor**. Variables cannot be reshaped after creation, only modified. Calling **tf.reshape** on a variable returns a new tensor. Variables can also be created from another variable, but the operation copies the underlying tensor. Variables do not share underlying data. **assign** can be used to update the variable by changing its data tensor. **assign_add** is another useful method of a variable that replicates the functionality of the += operator. Operations on tensors like **matmul** or **einsum** can also be applied to variables or to a combination of tensor and variable. Variable has a Boolean attribute called **trainable** that signifies if the variable is to be trained during backpropagation. Operations on variables are shown in Listing 2-6.

***Listing 2-6.*** Variables

```
1   import tensorflow as tf
2
3   tensor = tf.constant([[1, 2], [3, 4]])
4   variable = tf.Variable(tensor)
5   print(variable)
6   <tf.Variable 'Variable:0' shape=(2, 2) dtype=int32, numpy=
7   array([[1, 2],
8   [3, 4]])>
9
10  # return the index of highest element
11  print(tf.math.argmax(variable))
12
13  tf.Tensor([1 1], shape=(2,), dtype=int64)
14
15  print(tf.convert_to_tensor(variable))
16  tf.Tensor(
```

```
17   [[1 2]
18   [3 4]], shape=(2, 2), dtype=int32)
19
20   print(variable.assign([[1,2], [1, 1]]))
21   <tf.Variable 'UnreadVariable' shape=(2, 2) dtype=int32,
     numpy=
22   array([[1, 2],
23   [1, 1]])>
```

## 2.2  Graphs, Operations, and Functions

There are two modes of execution within TensorFlow: eager execution and graph execution. Eager mode of execution processes instructions as they occur in the code, while graph execution is delayed. Graph mode builds a dependency graph connecting the data represented as tensors (or variables) using operations and functions. After the graph is built, it is executed. Graph execution offers a few advantages over eager execution:

1. Graphs can be exported to files or executed in non-Python environments such as mobile devices.

2. Graphs can be compiled to speed up execution.

3. Nodes with static data and operations on those nodes can be precomputed.

4. Node values that are used multiple times can be cached.

5. Branches of the graph can be identified for parallel execution.

Operations in TensorFlow are represented using the **tf.Operation** class and can be used as a node. Operation nodes are created using one of the predefined operations such as **tf.matmul**, **tf.reduce_sum**, etc. To create a new operation, use the **tf.Operation** class. A few important operations are enumerated in the following. All of them can be accessed directly using the **tf.operation_name** syntax.

1. Operations defined in the **tf.math** library:

   - **tf.abs**: Calculates the absolute value of a tensor.

   - **tf.divide**: Divides two tensors.

   - **tf.maximum**: Returns the element-wise maximum of two tensors.

   - **tf.reduce_sum**: Calculates the sum of all tensor elements. It takes an optional axis argument to calculate the sum along that axis.

2. Operations defined in the **tf.linalg** library:

   (a). **tf.det**: Calculates the determinant of a square matrix

   (b). **tf.svd**: Calculates the SVD decomposition of a rectangular matrix provided as a tensor

   (c). **tf.trace**: Returns the trace of a tensor

Functions are defined using the **tf.function** method, passing the Python function as an argument. **tf.function** is a decorator that augments a Python function with attributes necessary for running it in a TensorFlow graph. A few examples of TensorFlow operations and functions are illustrated in Listing 2-7. Each TensorFlow function generates an internal graph from its arguments. By default, a TensorFlow function uses a graph execution model. To switch to eager execution mode, set **tf.config.run_functions_eagerly(True)**. Please note that the following output may not match output from another run because of random numbers used.

***Listing 2-7.*** TensorFlow Operations and Functions

```
1    import tensorflow as tf
2    import numpy as np
3
4    tensor = tf.constant(np.ones((3, 3), dtype=np.int32))
5
6    print(tensor)
7
8    <tf.Tensor: id=0, shape=(3, 3), dtype=int32, numpy=
9    array([[1, 1, 1],
10   [1, 1, 1],
11   [1, 1, 1]])>
12
13   print(tf.reduce_sum(tensor))
14   <tf.Tensor: id=2, shape=(), dtype=int32, numpy=9>
15
16   print(tf.reduce_sum(tensor, axis=1))
17   <tf.Tensor: id=4, shape=(3,), dtype=int32, numpy=
     array([3, 3, 3])>
18
19   @tf.function
20   def sigmoid_activation(inputs, weights, bias):
21       x = tf.matmul(inputs, weights) + bias
22       return tf.divide(1.0, 1 + tf.exp(-x))
23
24   inputs = tf.constant(np.ones((1, 3), dtype=np.float64))
25   weights = tf.Variable(np.random.random((3, 1)))
26   bias = tf.ones((1, 3), dtype=tf.float64)
```

```
27
28   print(sigmoid_activation(inputs, weights, bias))
29   <tf.Tensor: id=195, shape=(1, 3), dtype=float64,
     numpy=array([[0.89564016, 0.89564016, 0.89564016]])>
```

Code shown in Listing 2-8 sets the default execution mode to graph mode.

***Listing 2-8.*** Running TensorFlow Operations in Graph (Non-eager) Mode

```
1   import timeit
2
3   tf.config.experimental_run_functions_eagerly(False)
4   t1 = timeit.timeit(lambda: sigmoid_activation(inputs,
    weights, tf.constant(np.random.random((1, 3)))),
    number=1000)
5   print(t1)
6   0.7758807
```

# 2.3 Modules

TensorFlow uses the base class **tf.Module** to build layers and models. A module is a class that keeps track of its state using instance variables and can be called as a function. To achieve this, it must provide an implementation for the method __**call**__. This is illustrated in Listing 2-9. Due to the use of random numbers, output values may vary from those shown.

***Listing 2-9.*** Custom Module

```
1   import tensorflow as tf
2   import numpy as np
3
4
```

```
5   class ExampleModule(tf.Module):
6       def __init__(self, name=None):
7           super(ExampleModule, self).__init__(name=name)
8           self.weights = tf.Variable(np.random.random(5),
            name="weights")
9           self.const = tf.Variable(np.array([1.0]),
            dtype=tf.float64,
10          trainable=False, name="constant")
11
12      def __call__(self, x, *args, **kwargs):
13          return tf.matmul(x, self.weights[:, tf.newaxis]) +
            self.const[tf.newaxis, :]
14
15
16  em = ExampleModule()
17  x = tf.constant(np.ones((1, 5)), dtype=tf.float64)
18  print(em(x))
19
20
21  <tf.Tensor: id=24631, shape=(1, 1), dtype=float64,
    numpy=array([[2.45019464]])>
```

Module is the base class for both layers and models. It can be used as
a model, serving as a collection of layers. Module shown in Listing 2-10
defers the creation of weights for the first layer until inputs are provided.
Once input shape is known, it creates the tensors to store the weights.
Decorator **tf.function** can be added to the __call__ method to convert it to
a graph.

*Listing 2-10.* Module

```
1    import tensorflow as tf
2
3
4    class InferInputSizeModule(tf.Module):
5        def __init__(self, noutput, name=None):
6            super().__init__(name=name)
7            self.weights = None
8            self.noutput = noutput
9            self.bias = tf.Variable(tf.zeros([noutput]),
             name="bias")
10
11       def __call__(self, x, *args, **kwargs):
12           if self.weights is None:
13               self.weights = tf.Variable(tf.random.
                 normal([x.shape[-1], self.noutput]))
14
15           output = tf.matmul(x, self.weights) + self.bias
16           return tf.nn.sigmoid(output)
17
18   class SimpleModel(tf.Module):
19       def __init__(self, name=None):
20           super().__init__(name=name)
21
22           self.layer1 = InferInputSizeModule(noutput=4)
23           self.layer2 = InferInputSizeModule(noutput=1)
24
25       @tf.function
26       def __call__(self, x, *args, **kwargs):
27           x = self.layer1(x)
28           return self.layer2(x)
```

```
29
30    model = SimpleModel()
31    print(model(tf.ones((1, 10))))
32
33    <tf.Tensor: id=24700, shape=(1, 1), dtype=float32,
      numpy=array([[0.632286]], dtype=float32)>
```

Objects of type **tf.Module** can be saved to checkpoint files. Creating a checkpoint creates two files: one with module data and another containing metadata with extension **.index**. Saving a module to a checkpoint and loading it back from a checkpoint is illustrated in Listing 2-11.

***Listing 2-11.*** Checkpoint a Model

```
1     import tensorflow as tf
2
3     path = r"C:\temp\simplemodel"
4     checkpoint = tf.train.Checkpoint(model=model)
5     checkpoint.write(path)
6
7
8     model2 = SimpleModel()
9     model_orig = tf.train.Checkpoint(model=model2)
10    model_orig.restore(path)
```

# 2.4 Layers

Layers are objects with **tf.keras.layers.Layer** as the base class. The Keras library is used in TensorFlow for implementing layers and models. The **tf.keras.layers.Layer** class derives from the **tf.Module** class and has a method **call** in place of the __**call**__ method in **tf.Module**. There are several advantages to using Keras instead of **tf.Module**. For instance, training variables of nested Keras layers are automatically collected for training during backpropagation, whereas with **tf.Module**, variables have to be collected explicitly by the programmer. Additionally, one can provide an optional build method that gets called the first time Layer is invoked using the **call** method to initialize layer weights or other state variables based on input shape.

According to TensorFlow convention, input is always a two-dimensional or higher tensor. The first dimension indicates the batches. For example, if we have a set of N inputs, with each input comprised of one feature, input shape will be ($N$, 1). Notice how TensorFlow requires the first dimension to correspond to the number of batches. Similarly, the first dimension of output is the number of batches.

TensorFlow layers are derived from base class **tf.keras.layers.Layer**. A layer has the following noteworthy methods. For a full list, please check the TensorFlow API reference:

1. The **__init__(self)** method to initialize layer weights or other instance variables.

2. The **build(self, input_shape)** method is optional. When provided, it gets called the first time Layer is called with the **input_shape** parameter.

3. The **call(self, inputs, *args, **kwargs)** method takes the input and produces the output. This method takes two optional arguments listed in the following:

   - **training**: A Boolean argument if the call to Layer is made during the training period or prediction period. This argument may be used if the layer needs to do special work during training or prediction calls.

- **mask**: A Boolean tensor indicating some mask. For example, a layer could apply special logic to inputs if their batch number is present in the mask, or a recurrent neural network layer can use this to flag special timesteps.

4. The **get_config(self)** method returns a dictionary with layer configurations that need to be serialized when saving a checkpoint.

5. **weights** is a property of the **Layer** class and cannot be set in derived classes. Variables, that is, instances of type **Variable**, that are assigned as instance attributes become constituents of the **weights** property.

6. **trainable_weights** is also a property of the **Layer** class that contains trainable weights of this layer.

7. **add_loss**: Add additional losses like a regularization loss to the loss function.

8. **add_metric**: Add additional metrics for tracking training performance.

9. **get_weights**: Get all the weights – both trainable and non-trainable – of a layer as a list of numpy arrays.

10. **set_weights**: Set the weights of this layer to those provided in the list of numpy arrays. The structure of this list must be identical to the list returned by **get_weights**.

Sample code shown in Listing 2-12 creates a custom Keras layer that applies an upper bound of 0.9 on all its inputs. Before the first call to Layer, the **build** method has not been called, and **weights** is empty. After the first call to Layer, **weights** and **trainable_weights** properties have been initialized as seen from the output.

***Listing 2-12.*** Writing a Customized Layer

```
1   import tensorflow as tf
2   from tensorflow.keras.layers import Layer
3
4
5   class CustomDenseLayer(Layer):
6       def __init__(self, neurons):
7           super().__init__()
8           self.neurons = neurons
9
10      def build(self, input_shape):
11          # input_shape[-1] is the number of features for
            this layer
12          self.wt = tf.Variable(tf.random.normal((input_
            shape[-1], self.neurons), dtype=tf.float32),
13          trainable=True)
14          self.bias = tf.Variable(tf.zeros((self.neurons,),
            dtype=tf.float32),
15          trainable=True)
16          self.upperBound = tf.constant(0.9, dtype=tf.
            float32, shape=(input_shape[-1],))
17
18      def call(self, inputs):
19          return tf.matmul(tf.minimum(self.upperBound,
            inputs), self.wt) + self.bias
20
21
22  layer = CustomDenseLayer(5)
23  print(layer.weights)
24  print(layer.trainable_weights)
25
```

```
26   []
27   []
28
29   input = tf.random_normal_initializer(mean=0.5)
     (shape=(2, 5), dtype=tf.float32)
30   print(layer(inputs=input))
31
32   <tf.Tensor: id=171, shape=(2, 5), dtype=float32, numpy=
33   array([[-1.1098292 , -0.2773003 ,  0.24687909,  1.0952137 ,
     1.221024  ],
34   [-1.116677  , -0.4057744 ,  0.18726291,  1.0598873 ,
     1.3692323 ]],
35   dtype=float32)>
36
37   print(layer.weights)
38
39   [<tf.Variable 'custom_dense_layer_4/Variable:0' shape=(5, 5)
     dtype=float32, numpy=
40   array([[-1.3313855 , -0.7012864 , -1.003786  , -0.6224709 ,
     3.0700085 ],
41   [-0.1896328 ,  1.156029  ,  0.5904321 ,  0.20901136,
     -0.6205104 ],
42   [-0.13661204, -1.201732  , -0.08776241,  0.64640564,
     -0.9309348 ],
43   [-0.6379096 ,  0.43822217, -0.13019271,  0.4309327 ,
     0.8983831 ],
44   [ 0.03697195, -0.30708486,  1.1169728 ,  1.5509295 ,
     0.3927749 ]],
45   dtype=float32)>, <tf.Variable 'custom_dense_layer_4/
     Variable:0' shape=(5,) dtype=float32, numpy=
     array([0., 0., 0., 0., 0.], dtype=float32)>]
```

```
46
47   print(layer.trainable_weights)
48
49   [<tf.Variable 'custom_dense_layer_4/Variable:0' shape=(5, 5)
     dtype=float32, numpy=
50   array([[-1.3313855 , -0.7012864 , -1.003786  , -0.6224709 ,
     3.0700085 ],
51   [-0.1896328 ,  1.156029  ,  0.5904321 ,  0.20901136,
     -0.6205104 ],
52   [-0.13661204, -1.201732  , -0.08776241,  0.64640564,
     -0.9309348 ],
53   [-0.6379096 ,  0.43822217, -0.13019271,  0.4309327 ,
       0.8983831 ],
54   [ 0.03697195, -0.30708486,  1.1169728 ,  1.5509295 ,
       0.3927749 ]],
55   dtype=float32)>, <tf.Variable 'custom_dense_layer_4/
     Variable:0' shape=(5,) dtype=float32, numpy=
     array([0., 0., 0., 0., 0.], dtype=float32)>]
```

Keras layers also provide the ability to add loss functions like a
regularization loss to the overall loss function and to track additional metrics.

***Listing 2-13.*** Creating a Custom Layer for Lasso (L1) Regularization

```
1    import tensorflow as tf
2    from tensorflow.keras.layers import Layer
3
4
5    class LassoLossLayer(Layer):
6        def __init__(self, features, neurons):
7            super().__init__()
8            self.wt = tf.Variable(tf.random.normal((features,
                neurons), dtype=tf.float32),
```

```
9            trainable=True)
10           self.bias = tf.Variable(tf.zeros((neurons,),
             dtype=tf.float32),
11           trainable=True)
12           self.meanMetric = tf.keras.metrics.Mean()
13
14       def call(self, inputs):
15           # LASSO regularization loss
16           self.add_loss(tf.reduce_sum(tf.abs(self.wt)))
17           self.add_loss(tf.reduce_sum(tf.abs(self.bias)))
18           # metric to calculate mean of inputs
19           self.add_metric(self.meanMetric(inputs))
20           return tf.matmul(inputs, self.wt) + self.bias
```

In practice, one rarely needs to create custom layers. TensorFlow provides a range of layers useful in different neural networks. A few of them are described in the following. For a complete list, refer to the TensorFlow API:

- **Average**: Takes the average of inputs.

- **AveragePooling1D**: One-dimensional pooling layer used in convolutional neural networks. It takes pooling size and stride arguments. **AveragePooling2D** and **AveragePooling3D** layers are also available.

- **BatchNormalization**: Normalizes the input by subtracting the batch mean and dividing by the batch standard deviation during training. During prediction, when the **training** argument is **False**, uses a moving average of the mean and standard deviation computed using the values from the training phase and the current batch mean and standard deviation.

- **Conv1D**: One-dimensional convolution layer with provided number of filters (or number of channels), kernel size, and stride. Two- and three-dimensional convolution layers are also available.

- **Conv1DTranspose**: Deconvolution layer that produces the inverse of a convolution layer.

- **Dense**: A layer that connects all neurons in the layer to features (layer inputs).

- **Dropout**: Randomly sets the **rate** proportion of inputs to zero during training while scaling up the remaining inputs by $frac{1}{1-rate}$ so that the sum of inputs is unchanged. This is helpful for preventing overfitting. During prediction, this layer is a pass-through, sending the inputs as outputs.

- **Embedding**: This layer takes an input of dimension **input_dim** and returns a corresponding embedding of dimension **output_dim**. **input_dim** and **output_dim** are constructor arguments for this layer.

- **MaxPool1D**: Pool the inputs within the kernel, selecting the maximum value of input. This layer is useful in convolutional neural networks. Two- and three-dimensional max pooling layers are also available.

- **Softmax**: Softmax layer that computes $p_i = \dfrac{1}{1-e^{-x_i}}$ and returns the normalized probability $\dfrac{p_i}{\sum_j p_j}$ for a vector of inputs $x_i$. This layer has no trainable weights.

Layer's activation function can be provided as a constructor argument. If the activation function is omitted, the unit activation function is applied by default, that is, $y = W \cdot X$.

# 2.5 Models

TensorFlow models have **tf.keras.Model** as the base class, which in turn derives from the **tf.keras.layers.Layer** class. Models can serve as a collection of layers. For example, a sequential model is a collection of layers that applies the input to the first layer, passing its output to the second layer as input, and so on. Because models have **Layer** as a base class, all functionality of layers is available in models. Models can be saved as a checkpoint, deriving this functionality from the **tf.Module** base class. Models also have a method **save** to serialize the model to a file. A serialized model can be loaded using the **tf.keras.models.load_model** command.

An example of a customized sequential layer is shown in code Listing 2-14. The model has two layers: a dense layer with ReLU (rectified linear unit) activation and a softmax layer. As can be seen, the outputs from the softmax layer add to 1 for each row. Due to the use of random numbers, output values may vary from those shown.

***Listing 2-14.*** Writing a Customized Model

```
1    import tensorflow as tf
2    from tensorflow.keras import Model
3
4
5    class CustomSequentialModel(Model):
6        def __init__(self, name=None, **kwargs):
7            super().__init__(name, **kwargs)
8            self.layer2 = tf.keras.layers.Softmax()
9            self.layer1 = tf.keras.layers.Dense(10,
             activation=tf.keras.activations.relu)
10
11       def call(self, inputs, training=None, mask=None):
12           x = self.layer1(inputs)
```

```
13               return self.layer2(x)
14
15    model = CustomSequentialModel()
16    output = model(tf.random.normal((2, 10), dtype=tf.float32))
17    print(output)
18    tf.Tensor(
19    [[0.07642513 0.25438178 0.06848245 0.0847797  0.06848245
      0.10721327
20    0.06848245 0.07157873 0.10768385 0.09249022]
21    [0.0404469  0.0404469  0.0404469  0.0404469  0.0404469
      0.06400955
22    0.0404469  0.60652715 0.0404469  0.04633499]], shape=(2, 10),
      dtype=float32)
23
24    print(tf.reduce_sum(output, axis=1))
25    tf.Tensor([1.            0.99999994], shape=(2,),
      dtype=float32)
```

TensorFlow provides a sequential model **tf.keras.Sequential**. Layers are added to a sequential model using the **add** method. The first layer to a sequential model takes an optional argument **input_shape** specifying the number of features. If input shape for the first layer is not specified, the model must be built before compiling it. The **build** method of the model class takes input shape as argument. Before a model can be fitted to training data, it must be compiled, specifying the optimizer and loss function. Once fitted, the model can be used for making predictions. Usage of a sequential model is illustrated using an example shown in Listing 2-15. The code creates a sequential model comprised of three dense layers. It is then compiled and fitted to data using backpropagation. Once trained, it can be used for predicting.

A few important methods of the **tf.keras.Sequential** model class are listed in the following:

1.  **add**: Add a layer to the sequential model.

2.  **compile**: Compile the model. This step is required before the model can be trained. It specifies the optimizer used, loss function, metrics, and if it should run eagerly or in graph mode.

3.  **compute_loss**: Calculates the loss given the predicted outputs, the inputs, and the outputs using the loss function supplied to the model. If predicted outputs are not provided, the method first predicts the output using the inputs. Calculates the loss between predicted output and output.

4.  **evaluate**: Evaluate the model in prediction mode. Since this is not training mode, layers such as dropout layers behave accordingly.

5.  **fit**: Fit the model using provided inputs and outputs using backpropagation. It accepts optional arguments such as **batch_size** that specifies the number of samples used in each stochastic gradient step and **epochs** that specifies the number of optimization iterations. Returns a history object that can be used to track the evolution of loss and metrics over training epochs.

6.  **predict**: Predict the output from the model.

7.  **get_layer**: Retrieve a layer from the model using an index or name.

8.  **save**: Saves the model to a file.

9. **summary**: Prints a summary of input and output shapes and trainable parameters in each layer.

10. **to_json**: Saves the model to a JSON file.

Use of these APIs is illustrated using an example shown in Listing 2-15. In this code, data is generated by adding Gaussian white noise to function $4x + 2.5$. A model is fitted to the dataset using no regularization first, followed by using L2 regularization. Predicted results are plotted.

*Listing 2-15.* Sequential Model

```
1    import tensorflow as tf
2    import numpy as np
3    import matplotlib.pyplot as plt
4    import seaborn as sns
5    sns.set_theme(style="whitegrid")
6
7    # generate data
8    x = np.linspace(0, 5, 400, dtype=np.float32)  # 400 points
     spaced from 0 to 5
9    x = tf.constant(x)
10   y = 4*x + 2.5 + tf.random.truncated_normal((400,),
     dtype=tf.float32)
11   sns.scatterplot(x.numpy(), y.numpy())
12   plt.ylabel("y = 4x + 2.5 + noise")
13   plt.xlabel("x")
14   plt.show()
15
16   # create test and training data
17   x_train, y_train = x[0:350], y[0:350]
```

```
18   x_test, y_test = x[350:], y[350:]
19
20   # create the model
21   seq_model = tf.keras.Sequential()
22   seq_model.add(tf.keras.layers.Dense(5, input_shape=(1,)))
23   seq_model.add(tf.keras.layers.Dense(10, activation=tf.
     keras.activations.relu))
24   seq_model.add(tf.keras.layers.Dense(1))
25   print(seq_model.summary())
26
27   # Custom loss function with optional regularization
28   class Loss(tf.keras.losses.Loss):
29       def __init__(self, beta, weights):
30           super().__init__()
31           self.weights = weights
32           self.beta = beta
33
34       def call(self, y_true, y_pred):
35           reg_loss = 0
36           for i in range(len(self.weights)):
37               reg_loss += tf.reduce_mean(tf.square(self.
                 weights[i]))
38           return tf.reduce_mean(tf.square(y_pred - y_true))
             + self.beta * reg_loss
39
40   my_loss = Loss(0, seq_model.get_weights())
41
42   # compile the model
43   seq_model.compile(optimizer=tf.keras.optimizers.Adam(),
44                     loss=my_loss,
```

```
45                     metrics=[tf.keras.metrics.
                       MeanSquaredError()])
46
47   # fit the model to training data
48   history = seq_model.fit(x_train, y_train, batch_size=10,
     epochs=10)
49
50   # plot the history
51   plt.plot(history.history["mean_squared_error"],
     label="mean_squared_error")
52   plt.ylabel("Mean Square Error")
53   plt.xlabel("Epoch")
54   plt.show()
55
56   # predict unseen test data
57   y_pred = seq_model.predict(x_test)
58   plt.plot(x_test, y_test, '.', label="Test Data")
59   plt.plot(x_test, 4*x_test+2.5, label="Underlying Data")
60   plt.plot(x_test, y_pred.squeeze(), label="Predicted Values")
61   plt.legend()
62   plt.show()
63
64
65   Model: "sequential"
66
```

| Layer (type) | Output Shape | Param # |
|---|---|---|
| dense (Dense) | (None, 5) | 10 |
| dense_1 (Dense) | (None, 10) | 60 |
| dense_2 (Dense) | (None, 1) | 11 |

```
67   Layer (type)          Output Shape           Param #
68
69   dense (Dense)         (None, 5)              10
70
71   dense_1 (Dense)       (None, 10)             60
72
73   dense_2 (Dense)       (None, 1)              11
```

```
74
75    Total params: 81
76    Trainable params: 81
77    Non-trainable params: 0
78
79    None
```

The model is first fitted using no regularization, setting $\beta = 0$ in the argument to the loss function. Prediction results on testing data are shown in Figure 2-1. As can be observed, predicted values are very close to the underlying data-generating function, indicating good performance in testing data. Figure 2-2 shows the history of mean square error over the epochs. As can be seen from Figure 2-2, mean square error has converged.

Next, L2 regularization loss is introduced by setting $\beta = 0.05$ in the argument to the loss function. Prediction results are plotted in Figure 2-3. Regularization loss penalizes the higher value of the weight, forcing it down. As a result, predicted values are lower than the underlying data-generating function. Regularization is helpful in fitting a model to data with outliers. The testing data has no outliers in this example.

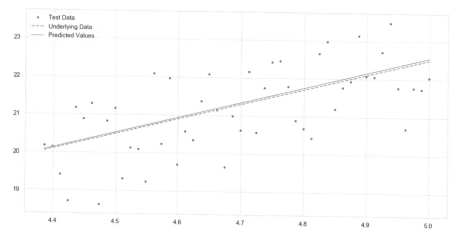

***Figure 2-1.*** *Predictions of the Model with No Regularization Against Underlying Data*

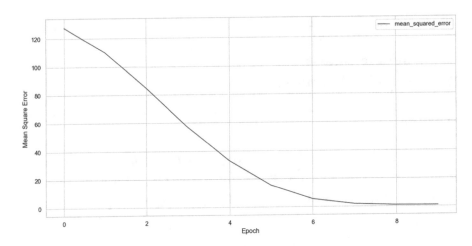

*Figure 2-2.* *History of Mean Square Error over Training Epochs*

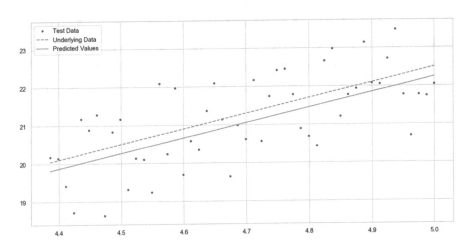

*Figure 2-3.* *Predictions of the Model with L2 Regularization Against Underlying Data*

# 2.6 Activation Functions

An activation function specifies the function applied to the dot product of neuron weights and inputs to determine the neuron's output. In equation 2.1, $g$ represents the activation:

$$y = g(W \cdot X + b) \tag{2.1}$$

TensorFlow has a number of predefined activation functions in module **tf.keras.activations**. A few of them are described in the following:

1. **ELU**: This is the exponential linear unit defined in **tf. keras.activations.elu**. Its activation function is illustrated in equation 2.2. $\alpha > 0$. For a large negative value of $x$, ELU saturates to a small negative value, $-\alpha$. ELUs help address the vanishing gradient problem because they do not saturate for large $x$:

$$y = \begin{cases} x \text{ if } x>0 \\ \alpha(e^x - 1) \text{ if } x<0 \end{cases} \tag{2.2}$$

2. **exponential**: Takes natural exponent $e^x$ of input.

3. **GELU**: Gaussian error linear unit that uses standard normal Gaussian CDF to calculate its output as shown in equation 2.3:

$$y = x \int_{-\infty}^{x} v \frac{1}{2\pi} e^{-\frac{v^2}{2}} dv \tag{2.3}$$

4. **ReLU**: Rectified linear unit activation produces $max(x, 0)$ as the output. It cuts off negative values at 0.

5.  The **LeakyReLU** activation function gives an output
    shown in equation 2.4. For positive values of $x$,
    it is identical to **ReLU**. Unlike **ReLU**, the output
    does not cut off to 0 for negative values of $x$. This
    helps avoid zero activation and zero gradients for
    negative values:

$$y = \begin{cases} \beta x \text{ if } x \geq 0 \\ \beta \alpha x \text{ if } x < 0 \end{cases} \quad (2.4)$$

6.  **SELU**: Scaled exponential linear unit activation
    scales the output of ELU activation by a scaling
    parameter $\beta$. Its output is shown in equation 2.5:

$$y = \begin{cases} \beta x \text{ if } x \geq 0 \\ \beta \alpha \left( e^x - 1 \right) \text{ if } x < 0 \end{cases} \quad (2.5)$$

7.  **sigmoid:** $y = \dfrac{1}{1 - e^{-x}}$ . This activation function
    saturates for large and small values of input $x$, giving
    rise to the vanishing gradient problem in deep
    neural networks and recurrent neural networks.

8.  **softmax**: Produces probability distribution from its
    inputs as shown in equation 2.6. Being a probability
    distribution, $\sum_i y_i = 1$ :

$$y_i = \frac{e^{x_i}}{\sum_j e^{x_j}} \quad (2.6)$$

9. **tanh**: Applies a hyperbolic tangent function as
   shown in equation 2.7 to produce output. Like
   the sigmoid function, it saturates for high and low
   values of input $x$:

$$y = \frac{e^x - e^{-x}}{e^x + e^{-x}} \qquad (2.7)$$

An activation function is provided as an argument to the layers object's constructor. Either the full name or a string can be used. TensorFlow keeps a mapping of strings to predefined activation functions. The two methods of specifying an activation function are shown in Listing 2-16. The advantage of using a fully qualified object name is that default arguments to the activation function can be changed.

*Listing 2-16.* Specifying an Activation Function

```
1   import tensorflow as tf
2   input = tf.random.normal((1, 5), dtype=tf.float32)
3   (input < 0).numpy().sum()
4   layer = tf.keras.layers.Dense(10, activation="relu",
    input_shape=(5,))
5   output = layer(input)
6   assert (output < 0).numpy().sum() == 0
7
8   layer2 = tf.keras.layers.Dense(10, activation=tf.keras.
    activations.relu, input_shape=(5,))
9   output2 = layer2(input)
10  assert (output2 < 0).numpy().sum() == 0
```

New activation functions can be added by defining a functor, that is, a class that can be instantiated and called using operator (), as shown in Listing 2-17. This example defines a new activation function $y = min\,(\alpha, x)$ where $\alpha$ is a configurable parameter set to 0.5. Inputs are all set to zero, giving $x = 0$ and output $y = \alpha$.

***Listing 2-17.*** Customizing an Activation Function

```
1    import tensorflow as tf
2
3    class MyActivation(object):
4        def __init__(self, alpha):
5            self.alpha = alpha
6
7        def __call__(self, x):
8            return tf.where(x < self.alpha, self.alpha, x)
9
10   layer = tf.keras.layers.Dense(1,
     activation=MyActivation(0.5), input_shape=(2,))
11   input = tf.constant([[0, 0]], dtype=tf.float32)
12   output = layer(input)
13   print(output)
14
15   tf.Tensor([[0.5]], shape=(1, 1), dtype=float32)
```

# 2.7  Loss Functions

A loss function defines a measure of difference between output and predicted output. Training a model involves adjusting the model's parameters to minimize the loss over a training dataset.

Loss functions have **tf.keras.losses.Loss** as their base class and override the method **call(y_true, y_pred)**. Predefined loss functions in TensorFlow can be found in module **tf.keras.losses**. A few loss functions from that module are described in the following:

1.  **BinaryCrossentropy**: Calculates loss between predicted labels and true labels in a binary (two-class) classification problem. Definition of the loss function is shown in equation 2.8. The constructor of this loss takes an argument **from_logits** indicating if the predicted outputs are true probabilities or un-normalized probabilities. The default value of **from_logits** is false. If true, $p_{class0} + p_{class1} = 1.0$ must hold. In equation 2.8, $I()$ is the indicator function. $p_{class0}(i)$ denotes the predicted probability of observation $i$ belonging to class 0:

$$L = -\sum_{i}\left[I\left(y_{true}\left(i\right)=0\right)\log\left(p_{class0}\left(i\right)\right)+I\left(y_{true}\left(i\right)=1\right)\log\left(1-p_{class1}\left(i\right)\right)\right] \quad (2.8)$$

2.  **CategoricalCrossentropy**: Calculates loss between predicted labels and true labels in a multiclass classification problem. Like its two-class cousin BinaryCrossentropy, it takes a **from_logits** argument indicating if the predicted outputs are true probabilities or un-normalized probabilities. Definition of this loss is shown in equation 2.9.

$p_{classj}(i)$ denotes the predicted probability of observation $i$ belonging to class $j$. True values must be provided as one-hot vectors:

$$L = -\sum_i \sum_{j \in classes} I\left(y_{true}(i) = j\right) \log\left(p_{classj}(i)\right) \qquad (2.9)$$

3. **CategoricalHinge**: This loss function is defined in equation 2.10. It is applicable to classification problems. $\tilde{p}_{classj}(i)$ depicts the normalized probability of observation $i$ belonging to class $j$. A model using this loss must produce normalized probabilities. This can be done by adding a softmax layer as the last layer:

$$L = \max\left[0, 1 + \sum_i \sum_{j \in classes} \left(I\left(y_{true}(i) \neq j\right)\tilde{p}_{classj}(i) - I\left(y_{true}(i) = j\right)\tilde{p}_{classj}(i)\right)\right] \quad (2.10)$$

Use of this loss function is illustrated using an example in Listing 2-18.

4. **CosineSimilarity**: Dot product of prediction and ground truth vectors normalized by L2 norm. The loss value is between –1 and 1, with –1 indicating perfect match between prediction and ground truth. Definition for this loss is shown in equation 2.11:

$$L = -\frac{1}{N}\sum_{i=1}^{N}\frac{\mathbf{y}(\mathbf{i})\cdot\hat{\mathbf{y}}(\mathbf{i})}{\|\mathbf{y}(\mathbf{i})\|^2\|\hat{\mathbf{y}}(\mathbf{i})\|^2} \qquad (2.11)$$

5. **Hinge**: Hinge loss is applicable to binary classification problems and is defined in equation 2.12. $y(i)$ is the ground truth and $\hat{y}(i)$ is the

prediction. This loss function assumes $y(i)$ is either 1 or –1 instead of the usual 0 and 1 binary class labels:

$$L = \max\left(0, 1 - y(i) - \hat{y}(i)\right) \tag{2.12}$$

6. **Huber**: This loss function is helpful in problems with outliers. Mean square error grows quadratically with deviation between actual and predicted values and can lead to poor convergence when outliers are present. Huber loss addresses this problem because it grows linearly for large deviations between actual and predicted values. This loss is defined in equation 2.13:

$$L \quad = \frac{1}{N}\sum_{i=1}^{N} L(i)$$

$$L(i) = \begin{cases} \frac{1}{2}\left(\hat{y}(i) - y(i)\right)^2 & \text{if } |\hat{y}(i) - y(i)| \le \delta \\ \delta\left(|\hat{y}(i) - y(i)| - \frac{1}{2}\delta\right) & \text{otherwise} \end{cases} \tag{2.13}$$

7. **MeanSquaredError**: This is perhaps the most widely used error function. It takes the mean of square deviations between observed and true values, as shown in equation 2.14:

$$L = \frac{1}{N}\sum_{i=1}^{N}\left(\hat{y}(i) - y(i)\right)^2 \tag{2.14}$$

8. **SparseCategoricalCrossentropy**: This loss function is functionally similar to **CategoricalCrossentropy**. It optimizes memory usage by relaxing the requirement of the **CategoricalCrossentropy** loss

function for its true values to be one-hot vectors.
Using **SparseCategoricalCrossentropy**, true values
(ground truth) should be integers indicating the
class number of the output. Indices begin from 0.

As an example, let us create a model to classify points into one of four
clusters. Each cluster has its mean x and y coordinates and a standard
deviation. Training data is constructed using 1000 random draws from
Gaussian distribution centered around each cluster's mean with that
cluster's standard deviation. This process gives a total of 4000 data points,
1000 points belonging to each cluster. The points are plotted in Figure 2-4.
The code is shown in Listing 2-18. The example uses five features: $x, y,$
$xy, x^2, y^2$.

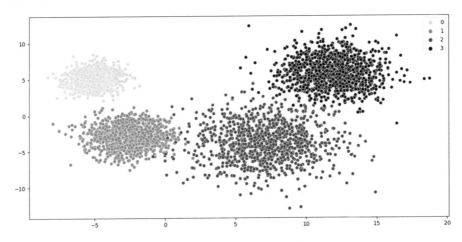

***Figure 2-4.*** *4000 Points Drawn from Four Clusters*

***Listing 2-18.*** Classifying a Point into One of Four Clusters

```
1    import tensorflow as tf
2    import numpy as np
3    import seaborn as sns
4    import matplotlib.pyplot as plt
```

```
5
6
7    class ClassifyCluster(object):
8        def __init__(self):
9            self.meansX = [-5, -2.5, 7, 12]
10           self.meansY = [5, -3, -4, 6]
11           self.stddevs = [1, 1.5, 2.7, 2]
12           self.nCluster = 4
13           self.nTraining = 4000
14           self.nTesting = 80
15           self.nFeature = 5
16           self.nnet = self.buildModel()
17
18       def trainModel(self):
19           # generate training data: 4 clusters with 1000
             points each
20           pts = self.nTraining // self.nCluster
21           randvals = np.random.standard_normal((pts, 2,
             self.nCluster)).astype(np.float32)
22           x = []
23           y = []
24
25           for i in range(4):
26               x.append(self.meansX[i] + self.
                 stddevs[i]*randvals[:, 0, i])
27               y.append(self.meansY[i] + self.
                 stddevs[i]*randvals[:, 1, i])
28
29           labels = np.repeat(np.arange(self.nCluster,
             dtype=np.int32), randvals.shape[0])
```

```
30          points_order = np.array(range(len(labels)),
            dtype=np.int32)
31          np.random.shuffle(points_order)
32
33          x_col = np.concatenate(x)
34          y_col = np.concatenate(y)
35
36          sns.scatterplot(x=x_col, y=y_col, hue=labels)
37          plt.show()
38
39          xy_col = np.multiply(x_col, y_col)
40          x2_col = np.multiply(x_col, x_col)
41          y2_col = np.multiply(y_col, y_col)
42
43          xy_data = np.concatenate((x_col[:, np.newaxis],
            y_col[:, np.newaxis], xy_col[:, np.newaxis],
44          x2_col[:, np.newaxis], y2_col[:, np.newaxis]),
            axis=1)
45          xy_data_tf = tf.constant(xy_data[points_order, :])
46          labels_tf = tf.constant(labels[points_order,
            np.newaxis])
47          history = self.nnet.fit(xy_data_tf, labels_tf,
            batch_size=20, epochs=15)
48          plt.plot(history.history["loss"])
49          plt.xticks(range(len(history.history["loss"])))
50          plt.xlabel("Epochs")
51          plt.ylabel("Categorical Crossentropy Loss")
52          plt.grid()
53          plt.show()
54
55          # find the accuracy on test data
```

```
56          result = self.nnet.predict(xy_data)
57          predicted_class = np.argmax(result, axis=1)
58          accuracy = (predicted_class == labels).sum() /
            float(labels.shape[0])
59          print(f"Model accuracy on training data =
            {accuracy}")
60
61          def buildModel(self):
62          # build the neural network model and train
63          nnet = tf.keras.models.Sequential()
64          nnet.add(tf.keras.layers.Dense(5, input_
            shape=(self.nFeature,)))
65          nnet.add(tf.keras.layers.Dense(15))
66          nnet.add(tf.keras.layers.Dense(4))
67          nnet.add(tf.keras.layers.Dense(4,
            activation="sigmoid"))
68          nnet.compile(optimizer=tf.keras.optimizers.
            Adam(learning_rate=0.005),
69          loss=tf.keras.losses.
            SparseCategoricalCrossentropy())
70          return nnet
71
72      def testModel(self):
73          # generate 80 points of testing data
74          randvals = np.random.standard_normal((self.
            nTesting, 2)).astype(np.float32)
75          test_labels = np.random.choice(self.nCluster,
            self.nTesting)
76          xy_test = np.ndarray((self.nTesting, self.
            nFeature), dtype=np.float32)
77          for i, label in enumerate(test_labels):
```

```
78              xy_test[i, 0] = self.meansX[label] + self.
                stddevs[label] * randvals[i, 0]
79              xy_test[i, 1] = self.meansY[label] + self.
                stddevs[label] * randvals[i, 1]
80              xy_test[i, 2] = xy_test[i, 0] * xy_test[i, 1]
81              xy_test[i, 3] = xy_test[i, 0] * xy_test[i, 0]
82              xy_test[i, 4] = xy_test[i, 1] * xy_test[i, 1]
83
84          result = self.nnet.predict(xy_test)
85          predicted_class = np.argmax(result, axis=1)
86          accuracy = (predicted_class == test_labels).sum() /
            float(test_labels.shape[0])
87          print(f"Model accuracy on testing data =
            {accuracy}")
88
89
90   if __name__ == '__main__':
91       classify = ClassifyCluster()
92       classify.trainModel()
93       classify.testModel()
```

A neural network model is constructed and trained using this test data, using the sparse categorical cross entropy loss function. The loss function plot over training epochs is shown in Figure 2-5. Accuracy on training data is around 99%. Testing data is constructed by randomly drawing 80 points from the clusters. Testing accuracy obtained is 97%.

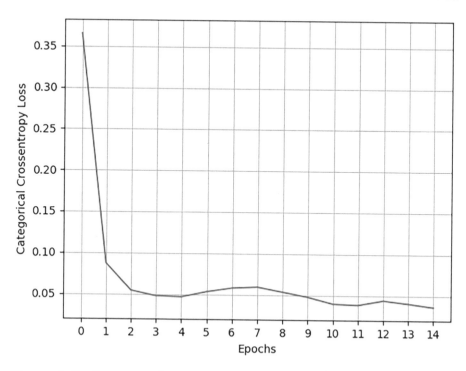

***Figure 2-5.*** *Categorical Cross Entropy Loss History During Training*

To create a new loss function, derive a class from base class **tf.keras. losses.Loss** and override the method **call(y_true, y_pred)**. An example is shown in Listing 2-19.

***Listing 2-19.*** Customizing a Loss Function

```
1    class CustomLoss(tf.keras.losses.Loss):
2        def call(self, y_true, y_pred):
3            return tf.reduce_mean(tf.abs(y_true - y_pred))
```

# 2.8 Metrics

Metrics are functions used to track the goodness of fit for a trained model. There are several ways to track the goodness of fit. For regression models, mean square error is a good metric. For models predicting a parametric probability density function, Kullback-Leibler divergence is a good candidate. For classification problems, a range of metrics are available depending upon the nature of data and prediction. Accuracy is the percentage of correct predictions. For binary classification problems, precision, recall, and F1 score are relevant metrics. Precision is the proportion of correct predictions for all the predictions in class "1". Recall is the proportion of correct predictions when the actual class is "1". Precision and recall typically move in opposite directions. Increasing the precision reduces the recall and vice versa. F1 score is a combination of precision and recall. To understand precision, recall, and F1 score, let us look at the confusion matrix of a binary classification problem shown in Figure 2-6.

**Prediction**

| Actual | | Positive | Negative | |
|---|---|---|---|---|
| | Positive | True Positive | False Negative | Recall = TP/(TP + FN) |
| | Negative | False Positive | True Negative | |
| | | Precision TP/(TP + FP) | | |

***Figure 2-6.*** *Confusion Matrix of a Binary Classification Problem*

Precision is defined as $\dfrac{TruePositive}{TruePositive + FalsePositive}$ , while recall is

defined as $\dfrac{TruePositive}{TruePositive + FalseNegative}$ . Accuracy is defined as

$\dfrac{TruePositive + TrueNegative}{AllData}$ . Precision measures how well a model

predicts a given class, while recall measures how well a model performs predicting a given outcome. For example, when using a cancer detector in a medical image, we want the detector to perform well for patients who have cancer, that is, predicting a given outcome. It is likely that the number of patients with cancer is small. In this case, true negatives may constitute the bulk of predictions. A model that assigns a negative outcome (i.e., no cancer) to all data points will achieve high accuracy and high precision but low recall. This is an example of imbalanced data within classes where one must use the appropriate metric.

TensorFlow has a number of predefined metrics available in module **tf.keras.metrics**. A few of them are listed in the following:

1. **AUC**: This represents area under the curve of ROC (receiver operating characteristic) curve for binary classification problems. ROC curve plots true positive rate (TPR) vs. false positive rate (FPR).

   $TPR = \dfrac{TP}{TP + FN}$ is also known as recall.

   $FPR = \dfrac{FP}{FP + TN}$ where TP, TN, FP, FN are true positive, true negative, false positive, and false negative, respectively. We want to increase true positive rate while reducing false positive rate. As the threshold in binary classification is increased, true positive rate increases because true positives increase, while false negatives decrease. This also increases false positive rate because false positives increase, while true negatives reduce or remain the same. The AUC metric is independent of threshold value that can be tweaked. A better model will have higher AUC. The AUC metric takes an optional argument **curve**. By setting it to "PR", area under the precision-recall curve can be calculated. Area under the precision-recall curve measures precision vs. recall performance of a binary classifier. In order to understand the plots, let us consider a binary classification problem shown in Figure 2-7. Data points shown with "–" are negatives, while data points shown with "+" are positives. An example distribution of data is shown in Figure 2-7. The dotted vertical line shows the classification threshold: points to its left are classified as negative, and points to its right or on it are classified as positive. Threshold is increased from 0 to 1. With the threshold line at 0, all points are classified as positive. This results in all positive points getting classified correctly. Since there are no negative label predictions, false negatives = 0 and true negatives = 0. This gives high recall (or true positive rate) = 1, high false positive rate = 1, but low precision = $\dfrac{TP}{N}$ where $N$ is the number of points and low false negative. This point is located in the top-right corner of the AUC-ROC curve in Figure 2-8 and the top left of the AUC-PR curve in Figure 2-9. As threshold moves to the right, more negative points are classified correctly, increasing true negative and false negative. $TP + FN$ remains constant, equal to the number of positive samples. Similarly, $FP + TN$ remains constant, equal to the number of negative samples. The number of true positives reduces, causing recall to fall. $TP + FP$ falls faster than $TP$ because $FP$ is also falling. This causes precision $= frac{TP}{TP + FP}$ to increase, leading to movement toward the lower-right corner of the AUC-PR curve in Figure 2-9. False positives reduce in number with $FP + TN$ remaining the same, causing false positive rate to fall. This leads to movement toward the lower-left corner of the AUC-ROC curve.

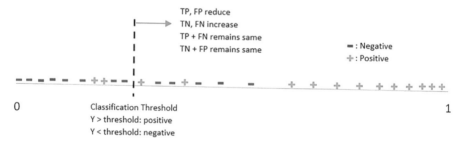

**Figure 2-7.** *Binary Classification: Changing Precision, Recall, and FPR as Threshold Moves*

2. **Accuracy**: This metric represents the proportion of data items classified correctly.

3. **BinaryIoU**: The binary intersection-over-union metric is defined as shown in equation 2.15. Its constructor takes two optional arguments: **target_class_ids** indicating the labels of the two classes in actual output and **threshold** that applies to

predicted values and considers values falling below it as one class and those falling on or above it as second class:

$$\text{Binary IoU} = \frac{TP}{TP + FP + FN} \qquad (2.15)$$

4. **CategoricalCrossentropy**: Similar to the categorical cross entropy loss function described in the previous section.

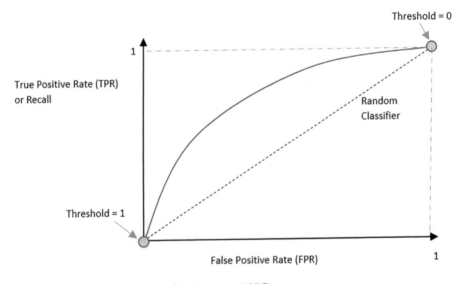

**Figure 2-8.**  *Area Under the Curve, AUC*

5. **FalsePositives**: This metric is equal to the number of false positive data points.

6. **KLDivergence**: Calculates Kullback-Leibler divergence as $y \log\left(\dfrac{y}{\hat{y}}\right)$ where $y$ is the known output and $\hat{y}$ is the predicted output. Sample code showing a calculation is illustrated in Listing 2-20.

***Listing 2-20.*** Kullback-Leibler Divergence Metric

```
1       import tensorflow as tf
2
3       metric = tf.keras.metrics.KLDivergence()
4       metric.update_state([[1, 0], [0, 1]], [[0.3, 0.7],
        [0.5, 0.5]])
5       print(metric.result().numpy())
6
7       0.94855845
```

   7.  **MeanSquaredError**: Mean square error between
      actual and predicted values.

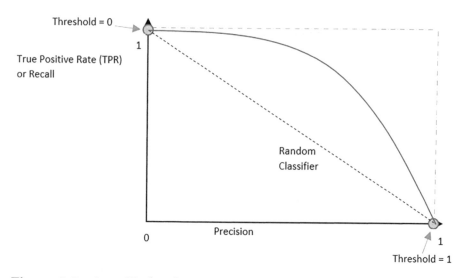

***Figure 2-9.*** *Area Under the Precision-Recall Curve, AUC-PR*

   8.  **Precision**: Precision is defined as $\dfrac{TP}{TP+FP}$ and
      measures the accuracy of predicting true. It is also
      called positive predictive value.

9. **PrecisionAtRecall**: Calculates highest precision when recall is $\geq$ a threshold.

10. **Recall**: Recall is defined as $\dfrac{TP}{TP + FN}$ and is also known as sensitivity or true positive rate (TPR) for a binary classification problem. For a multiclass classification problem, recall can be defined separately for each label by converting the multiclass problem to a binary class problem for each label (output class is that label or not).

Let us look at an example of predicting fraudulent credit card transactions. The dataset is hosted on Kaggle and has a total of 284,807 transactions, out of which 0.17% or 492 transactions are fraudulent. This is an example of an imbalanced dataset. Financial institutions have an incentive to detect fraud and also to avoid flagging authentic transactions as fraud in order to ensure customer satisfaction, that is, the predictor should have high recall and high precision. The data is downloaded from Kaggle's credit card fraud detection dataset (Kaggle, 2022). In order to safeguard data privacy, dataset columns have been anonymized and are reported as "V1" through "V28". "Amount" denotes the transaction amount, and "Class" denotes if the transaction is legitimate (0) or fraud (1).

This example illustrates a few general recommendations for effective neural network modeling:

1. **Normalize the inputs** so that they are neither too high nor too low. Machine learning models learn` faster if the inputs are comparable. For example, if all input features are in the [–1, 1] range, training will be faster than for the case where input features have widely dispersed ranges. For a standard Gaussian distribution, 95% of probability density lies between

[−1.95, 1.95]. Normalizing input features using

$\dfrac{x - \mu}{2\sigma}$ where $x$ is the feature value and $(\mu, \sigma)$ are the

mean and standard deviation of the feature value over the training dataset gives a simple method of normalizing inputs. Outliers present a challenge; outliers will show up as data points with large normalized feature values. A few solutions to deal with outliers are

- Using a loss function that is robust to outliers, such as Huber loss

- Applying L1 or L2 regularization that prevents model weights from becoming too large in absolute value

- Using input feature normalization and clipping the value of outliers

Only the training dataset should be used to calculate these normalizing hyper-parameters. During testing, hyper-parameter values should be frozen. For example, the **BatchNormalization** layer in TensorFlow normalizes inputs using equation 2.16. $\alpha$ represents momentum for the moving average:

$$
\begin{aligned}
x_{normalized} &= \gamma \frac{x - \mu}{\sigma + \epsilon} + \beta \\
\mu &\leftarrow \alpha\mu + (1-\alpha)\mu_{batch}(x) \\
\sigma^2 &\leftarrow \alpha\sigma^2 + (1-\alpha)\sigma^2_{batch}(x)
\end{aligned}
\tag{2.16}
$$

In the implementation shown in Listing 2-21, input features have been normalized using $\frac{x-\mu}{2\sigma}$. The first two input features are plotted after normalization for legitimate and fraudulent transactions in the training dataset in Figures 2-10 to 2-13.

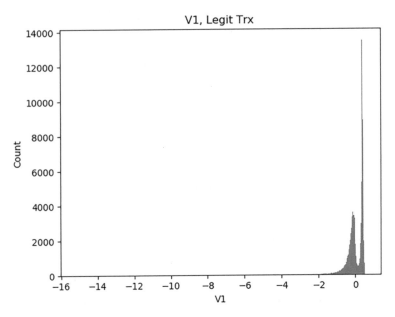

***Figure 2-10.*** *Normalized Feature V1 in Legitimate Transactions, Training Dataset*

As can be seen from Figures 2-10 and 2-11, normalized V1 is more than –2 for legitimate transactions, whereas it can be as small as –8 for fraudulent transactions. This indicates feature V1 will likely be important in classification.

Features that show identical distributions between classes (legitimate and fraudulent transactions) can be dropped from input because they will have marginal predictive power.

2.  Reduce the number of input features to those that are necessary for classification. Examining the distribution of normalized inputs against output can be helpful. Inputs that have identical distribution across all output classes can be dropped.
    **Parsimonious models** do not have surplusage; they only have features relevant for the classification task. This improves model training because there are fewer model parameters to learn, avoids overfitting, and improves performance on a testing dataset. Increasing the number of redundant input features will add more free parameters, leading to overfitting on a training dataset and poor performance on a testing dataset.

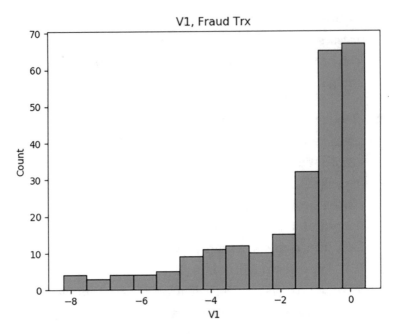

***Figure 2-11.*** *Normalized Feature V1 in Fraudulent Transactions, Training Dataset*

3. Training-testing split of 80%-20% is used.

4. **Use model checkpointing** to promote reproducibility of results across different runs. Without using a checkpoint, TensorFlow will randomly initialize the network weights, and results will differ across runs for initial training epochs before convergence.

5. **Initialize output layer bias**: Output layer bias should be initialized to a value that makes network output close to average output observed in testing data. For example, a neural network that predicts a binary class label and has sigmoid function activation in the final layer can be initialized so as to

set the final layer bias using equation 2.17. $p$ denotes the number of data points in the positive class, while $n$ denotes the data points in the negative class:

$$\frac{1}{1+e^{-bias}} = \frac{p}{p+n}$$

$$bias = \ln\left(\frac{p}{n}\right)$$

(2.17)

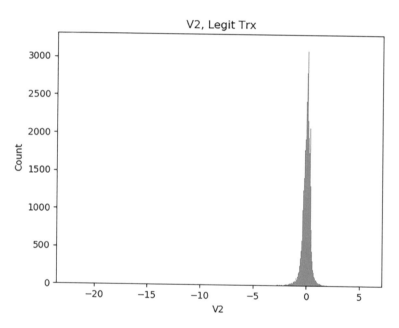

**Figure 2-12.** *Normalized Feature V2 in Legitimate Transactions, Training Dataset*

6. Ensure **data distribution is consistent across the training dataset**. For example, if one is building a neural network model to predict income based on features such as age, education, and years in the workforce and profession, one must ensure that

57

entire data is from identical probability distribution.
Sex and country may be important determinants
of income. If those two features are excluded and
the dataset contains data from different sexes
and countries, data will likely be from different
distributions, leading to a poor model performance.

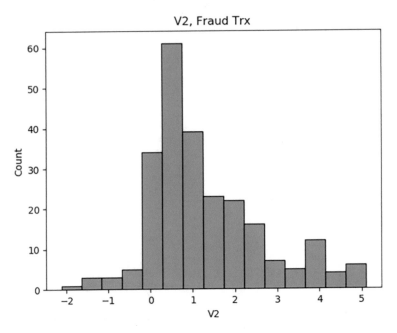

**Figure 2-13.** *Normalized Feature V2 in Fraudulent Transactions,*
*Training Dataset*

7. **Avoid information leakage or in-sample bias**.
   This will lead to the model producing deceptively
   good predictions in the training dataset but failing
   to live up to expectations in the testing dataset.
   Information leakage can occur in several ways.
   Hyper-parameter selection must use the training
   dataset. Feature normalization should use values
   from the testing dataset.

8.  For ***imbalanced class data, identify the primary
    objective of the model.*** If one plans to achieve high
    accuracy and high precision at the cost of lower
    recall, training using cross entropy loss should work.
    If high recall value is required, model precision will
    typically reduce. As can be seen from precision-
    recall curves in Figure 2-14, there is a trade-off
    between these two metrics. Figure 2-14 also shows
    a similar model performance in training and
    validation datasets.

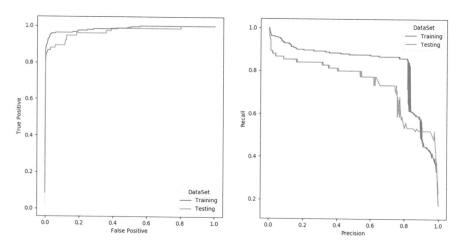

***Figure 2-14.***  *Area Under the Curve (AUC) for ROC and Precision-
Recall Curves*

To increase recall in an imbalanced class dataset,
one must either assign class weights to mitigate
class imbalance or perform resampling by drawing
additional samples from the class with lower
frequency. Both of these methods are described in
the following.

- Class weights for the classes can be set using
  equation 2.18. A class with less data will get a
  higher weight, and the one with more data will get
  a lower weight. The weights are used as multipliers
  with loss contributions from the two classes. Class
  weights need to be provided as a dictionary to the
  fit method:

$$cw_+ = \frac{n_-}{n_+}$$

$$cw_- = \frac{n_+}{n_-}$$

(2.18)

Figure 2-15 shows area under the curve for ROC and
precision-recall curves for training and validation
datasets. Area under the precision-recall curve is
higher using class weights, with higher precision
and recall values.

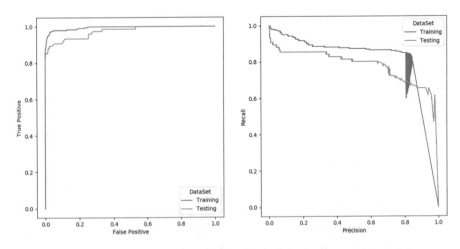

**Figure 2-15.**  *Area Under the Curve (AUC) for ROC and Precision-Recall Curves with Class Reweighing*

- Upsample the class with lower frequency. Upsampling augments data points of the less frequent class, thereby removing the class imbalance. Figure 2-16 shows area under ROC and precision-recall curves for training and validation datasets using upsampling for the positive class (fraudulent transactions). As with class reweighing, resampling improves precision and recall. An example of upsampling is shown in the code in Listing 2-21.

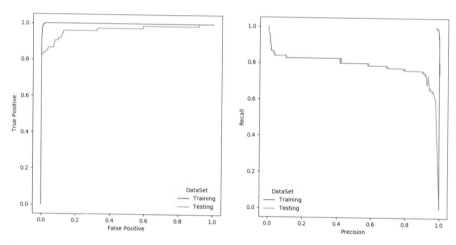

***Figure 2-16.*** *Area Under the Curve (AUC) for ROC and Precision-Recall Curves with Resampling*

9. Plot confusion matrices for testing data to visualize the accuracy, precision, and recall of the model in out-of-sample testing data. This gives an indication of the model's performance. Confusion matrices using a threshold value of 0.5 are plotted for the original model, the model with class weights, and

the model with resampling in Figures 2-17, 2-18, and 2-19, respectively. Because the model produces a probability, we need to provide a threshold. A probability value below the threshold is assigned the negative class and above the threshold is assigned the positive class. The original version of the model (Figure 2-17) has recall $= \dfrac{45}{30+45} = 60\%$,

precision $= \dfrac{45}{10+45} \approx 81.8\%$, and accuracy

$= \dfrac{56876+45}{56876+45+10+30} \approx 99.9\%$. Out-of-sample accuracy is good, but recall is low. With class weighing (Figure 2-18), out-of-sample recall

increases to $\dfrac{66}{9+66} \approx 88\%$, but precision

drops to $\dfrac{66}{1492+66} \approx 4.24\%$. Accuracy falls to

$\dfrac{55394+66}{55394+66+1492+9} \approx 97.4\%$. This indicates the

precision-recall trade-off observed earlier. With resampling (Figure 2-19), out-of-sample recall is

$\dfrac{63}{12+63} \approx 84\%$, and precision is $\dfrac{63}{1182+63} \approx 5.06\%$.

Accuracy is $\dfrac{55394+66}{55394+66+1492+9} \approx 97.9\%$.

10. Plot the metrics and loss function to ensure model training has converged. As seen from Figures 2-20, 2-21, and 2-22 for the original model, the model with class weighing, and the model with resampling, metrics and loss functions have converged by 20 epochs for training and validation datasets.

***Listing 2-21.*** Predicting Credit Card Fraud

```
1    import numpy as np
2    import pandas as pd
3    import tensorflow as tf
4    import os
5    import matplotlib.pyplot as plt
6    import seaborn as sns
7    from sklearn.metrics import confusion_matrix, roc_curve,
     precision_recall_curve
8    from typing import List
9
10
11   class FraudDetector(object):
12       def __init__(self, inputdir, checkpoint=True):
13           self.inputDir = inputdir
14           self.featureColumns = ["Amount"] + [f"V{i}" for i
             in range(1, 29)]
15           self.resultColumn = "IsFraud"
16           self.normalize = {}
17           self.dataDf = pd.read_csv(os.path.join
             (self.inputDir, "creditcard.csv"))
18           self.dataDf.loc[:, "IsFraud"] = (self.dataDf.
             Class == 1)
19           self.batchSize = 1024
20           self.nEpoch = 20
21           self.testingDataSize = 0.2
22           self.validateDataSize = 0.1
23           self.trainDf, self.validateDf, self.testDf =
             self.splitTrainValTestData()
24           self.calcNormalizingConst(self.trainDf)
```

```
25      self.metrics = [tf.keras.metrics.
        TruePositives(name="TP"),
26                      tf.keras.metrics.False
                        Negatives(name="FN"),
27                      tf.keras.metrics.False
                        Positives(name="FP"),
28                      tf.keras.metrics.Binary
                        Accuracy(name="Acc"),
29                      tf.keras.metrics.Precision
                        (name="Prec"),
30                      tf.keras.metrics.Recall
                        (name="Recall"),
31                      tf.keras.metrics.AUC
                        (name="aucroc"),
32                      tf.keras.metrics.AUC(curve="PR",
                        name="aucpr")]
33      self.plotMetrics = ["TP", "FN", "FP", "Acc",
        "Prec", "Recall", "aucroc", "aucpr"]
34      self.plotMetricsLabels = ["True Pos", "False
        Neg", "False Pos", "Accuracy", "Precision",
        "Recall", "AUC ROC", "AUC PR"]
35      self.checkpoint = checkpoint
36      self.nnet = self.model()
37      if checkpoint:
38          self.nnet = self.checkpointModel(self.nnet)
39
40  def plotNormalizedVars(self):
41      data = {}
42      for column in self.featureColumns:
43          mean, sd = self.normalize[column]
44          transformedCol = (self.trainDf.loc[:,
            column].values - mean) / sd
```

```
45              data[column] = transformedCol
46
47          data[self.resultColumn] = self.trainDf.loc[:,
            self.resultColumn].values
48          df = pd.DataFrame(data)
49          for plotcol in self.featureColumns:
50              sns.histplot(data=df.loc[~df.loc[:,
                self.resultColumn], :], x=plotcol).
                set(title=plotcol + ", Legit Trx")
51              plt.show()
52              sns.histplot(data=df.loc[df.loc[:,
                self.resultColumn], :], x=plotcol).
                set(title=plotcol + ", Fraud Trx")
53              plt.show()
54
55      def plotConfusionMatrix(self, labels: np.ndarray,
        predictions: np.ndarray, thresh=0.5) -> None:
56          cm = confusion_matrix(labels, predictions >
            thresh)
57          sns.heatmap(cm, annot=True, fmt="d",
            linewidths=0.25)
58          plt.xticks([0, 1, 2])
59          plt.yticks([0, 1, 2])
60          plt.title(f"Confusion Matrix,
            Threshold={thresh}")
61          plt.ylabel('Actual')
62          plt.xlabel('Predicted')
63          plt.show()
64
65      def plotAUC(self, labels: List[np.ndarray],
        predictions: List[np.ndarray]) -> None:
```

```
66    fp_train, tp_train, other = roc_curve(labels[0],
      predictions[0])
67    df_train = pd.DataFrame({'False Positive':
      fp_train, 'True Positive': tp_train, "DataSet":
      ["Training"] * len(fp_train)})
68    fp_test, tp_test, other = roc_curve(labels[1],
      predictions[1])
69    df_test = pd.DataFrame({'False Positive':
      fp_test, 'True Positive': tp_test, "DataSet":
      ["Testing"] * len(fp_test)})
70    df = pd.concat((df_train, df_test), axis=0,
      ignore_index=True)
71
72    axs = plt.subplot(1, 2, 1)
73    sns.lineplot(x="False Positive", y="True
      Positive", data=df, hue="DataSet", ax=axs)
74
75    precision_train, recall_train, other = precision_
      recall_curve(labels[0], predictions[0])
76    df_train = pd.DataFrame({'Precision': precision_
      train, 'Recall': recall_train, "DataSet":
      ["Training"] * len(precision_train)})
77    precision_test, recall_test, other = precision_
      recall_curve(labels[1], predictions[1])
78    df_test = pd.DataFrame({'Precision': precision_
      test, 'Recall': recall_test, "DataSet":
      ["Testing"] * len(precision_test)})
79    df = pd.concat((df_train, df_test), axis=0,
      ignore_index=True)
80
81    axs = plt.subplot(1, 2, 2)
```

```
82      sns.lineplot(x="Precision", y="Recall", data=df,
        hue="DataSet", ax=axs)
83      plt.show()
84
85  def testTrainSplit(self, df, test_size):
86      ntest = int(test_size * df.shape[0])
87      ntrain = df.shape[0] - ntest
88      return df.loc[0:ntrain, :].reset_
        index(drop=True), df.loc[ntrain:, :].reset_
        index(drop=True)
89
90  def splitTrainValTestData(self):
91      """ Returns training, validation and testing
        datasets as dataframes """
92      train, test = self.testTrainSplit(self.dataDf,
        test_size=self.testingDataSize)
93      train, validation = self.testTrainSplit(train,
        test_size=self.validateDataSize/self.
        testingDataSize)
94      return train, validation, test
95
96  def model(self):
97      npos = self.trainDf.loc[:, self.
        resultColumn].sum()
98      nneg = self.trainDf.shape[0] - npos
99      initBias = tf.keras.initializers.Constant(np.
        log(npos/float(nneg)))
100     nnet = tf.keras.models.Sequential()
101     nnet.add(tf.keras.layers.Dense(20,
        activation="relu", input_shape=(len(self.
        featureColumns),)))
```

```
102            nnet.add(tf.keras.layers.Dropout(0.2))
103            nnet.add(tf.keras.layers.Dense(1,
               activation="sigmoid", bias_initializer=initBias))
104            nnet.compile(optimizer=tf.keras.
               optimizers.Adam(),
105                         loss=tf.keras.losses.
                            BinaryCrossentropy(),
106                         metrics=self.metrics)
107            return nnet
108
109        def checkpointModel(self, nnet):
110            checkpointFile = os.path.join(self.inputDir,
               "checkpoint_init_wt")
111            if not os.path.exists(checkpointFile):
112                nnet.predict(np.ones((20, len(self.
                   featureColumns)), dtype=np.float32))
113                tf.keras.models.save_model(nnet,
                   checkpointFile, overwrite=False)
114            else:
115                nnet = tf.keras.models.load_
                   model(checkpointFile)
116            return nnet
117
118        def resampleData(self, trainFeatures, trainClass):
119            posData = trainClass[:, 0]
120            npos = posData.sum()
121            nneg = posData.shape[0] - npos
122            ids = np.where(posData)[0]
123            choice = np.random.choice(ids, nneg)
124            resample = np.concatenate((choice,
               np.where(~posData)[0]))
```

```
125         np.random.shuffle(resample)
126         trainFeatures = trainFeatures[resample, :]
127         trainClass = trainClass[resample, :]
128         return trainFeatures, trainClass
129
130     def trainModelAndPredict(self, useClassWeights=False,
        resamplePosData=False):
131         cols = []
132         validationCols = []
133         testCols = []
134         for column in self.featureColumns:
135             mean, sd = self.normalize[column]
136             transformedCol = (self.trainDf.loc[:,
                column].values - mean) / sd
137             cols.append(transformedCol[:, np.newaxis])
138             valCol = (self.validateDf.loc[:, column].
                values - mean) / sd
139             validationCols.append(valCol[:, np.newaxis])
140             testCol = (self.testDf.loc[:, column].
                values - mean) / sd
141             testCols.append(testCol[:, np.newaxis])
142         trainFeatures = np.concatenate(cols, axis=1)
143         validationFeatures = np.concatenate
            (validationCols, axis=1)
144         trainClass = self.trainDf.loc[:, self.
            resultColumn].values[:, np.newaxis]
145         validationClass = self.validateDf.loc[:, self.
            resultColumn].values[:, np.newaxis]
146         testFeatures = np.concatenate(testCols, axis=1)
147         testClass = self.testDf.loc[:, self.
            resultColumn].values
```

```
148         classWts = None
149         assert not (useClassWeights and resamplePosData),
            "useClassWeights and resamplePosData cannot both
            be True"
150         if useClassWeights:
151             npos = trainClass.sum()
152             nneg = trainClass.shape[0] - npos
153             classWts = {True: (1.0/npos) * trainClass.
                shape[0]/2.0,
154                         False: (1.0/nneg) * trainClass.
                            shape[0]/2.0}
155         if resamplePosData:
156             trainFeatures, trainClass = self.
                resampleData(trainFeatures, trainClass)
157         history = self.nnet.fit(trainFeatures,
            trainClass, batch_size=self.batchSize,
            epochs=self.nEpoch,
158                                 validation_
                                    data=(validationFeatures,
                                    validationClass),
159                                 class_weight=classWts)
160         self.plotHistory(history)
161         resTrain = self.nnet.predict(trainFeatures)
162         res = self.nnet.predict(testFeatures)
163         self.plotConfusionMatrix(testClass, res[:, 0])
164         labels = [trainClass[:, 0], testClass]
165         predic = [resTrain[:, 0], res[:, 0]]
166         self.plotAUC(labels, predic)
167
168     def plotHistory(self, history):
169         for n, metric in enumerate(self.plotMetrics):
```

```
170              plt.subplot(4, 2, n+1)
171              plt.plot(history.epoch, history.
                 history[metric], label="Training")
172              plt.plot(history.epoch, history.
                 history[f"val_{metric}"], linestyle='--',
                 label="Validation")
173              plt.xlabel("Epoch")
174              plt.ylabel(self.plotMetricsLabels[n])
175              plt.legend()
176          plt.show()
177
178      def calcNormalizingConst(self, testDf):
179          for column in self.featureColumns:
180              mean, sd = np.mean(testDf.loc[:, column].
                 values), np.std(testDf.loc[:, column].values)
181              self.normalize[column] = (mean, 2*sd)
182
183          self.plotNormalizedVars()
184
185
186  def main():
187      fdetect = FraudDetector(r"C:\prog\cygwin\home\
         samit_000\RLPy\data\book", True)
188      fdetect.trainModelAndPredict()
189      # use class weights
190      fdetect.trainModelAndPredict(useClassWeights=True)
191      # use resampling of positive class data
192      fdetect.trainModelAndPredict(resamplePosData=True)
193
194
195  if __name__ == "__main__":
196      main()
```

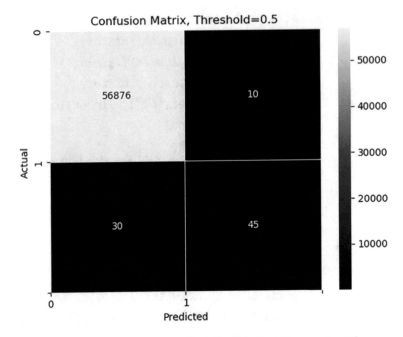

***Figure 2-17.*** *Confusion Matrix for the Testing Dataset with Theshold 0.5*

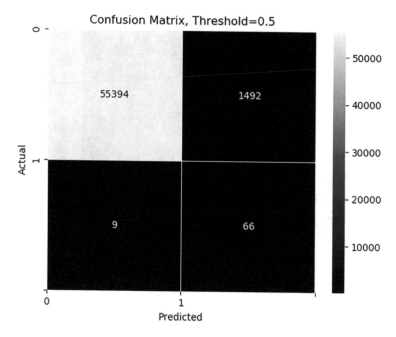

*Figure 2-18.* *Confusion Matrix for the Testing Dataset with Class Reweighing*

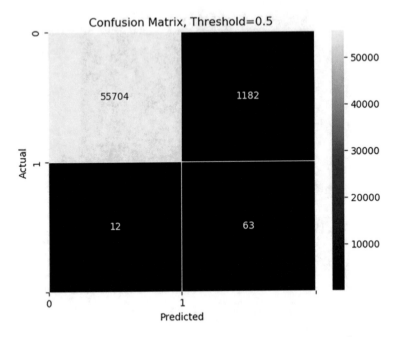

**Figure 2-19.** *Confusion Matrix for the Testing Dataset with Resampling*

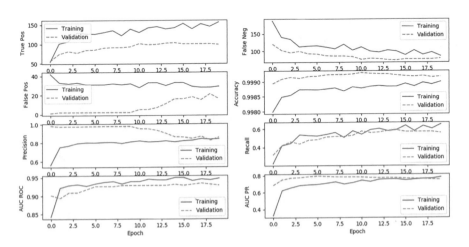

**Figure 2-20.** *Metrics and Loss Function Convergence in 20 Epochs*

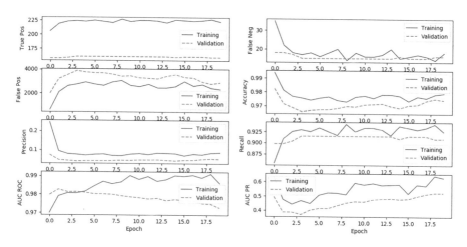

***Figure 2-21.***  *Metrics and Loss Function Convergence in 20 Epochs with Class Reweighing*

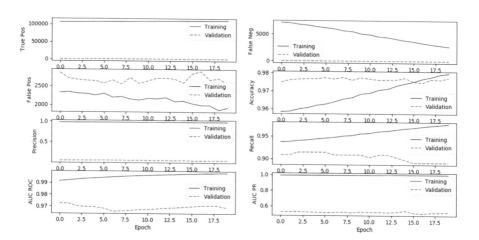

***Figure 2-22.***  *Metrics and Loss Function Convergence in 20 Epochs with Resampling*

In order to define a new metric, derive a class from **tf.keras.metric. Metric** and provide implementation of methods __init__, **update_state**, and **result**. F1 score is a metric composed of precision and recall. Precision and recall typically move in opposite directions – improving one reduces the other. This metric incorporates both of these metrics into one measure and is defined as shown in equation 2.19. An implementation of this metric is shown in Listing 2-22:

$$F_1 = \frac{2}{\dfrac{1}{recall} + \dfrac{1}{precision}} \tag{2.19}$$

***Listing 2-22.*** Creating the F1 Score Metric

```
1    import tensorflow as tf
2
3
4    class F1Score(tf.keras.metrics.Metric):
5        def __init__(self, thresholds=0.5, name="F1Score",
         **kwargs):
6            super().__init__(name=name, **kwargs)
7            self.recall = tf.keras.metrics.
             Recall(thresholds=thresholds)
8            self.precision = tf.keras.metrics.Precision(thresh
             olds=thresholds)
9            self.f1score = self.add_weight(name="f1",
             initializer='zeros')
10
11       def update_state(self, y_true, y_pred, sample_
         weights=None):
12           self.recall.update_state(y_true, y_pred, sample_
             weights)
```

```
13              self.precision.update_state(y_true, y_pred,
                sample_weights)
14              self.f1score.assign_add(1.0/(1.0/self.recall.
                result() + 1.0/self.precision.result()))
15
16      def result(self):
17              return self.f1score
18
19   f1 = F1Score()
20   f1.update_state([[0], [1], [1]], [[1], [0], [1]])
21   print(f1.result().numpy())
22
23   0.25
```

# 2.9 Optimizers

Optimizers are classes used to perform gradient descent in training neural network models using backpropagation. Stochastic gradient descent is used to search for a local optimum (minimum) of a loss function using a randomly drawn batch of inputs. Optimizers perform gradient descent and calculate change in network parameters. Speed of convergence can vary a lot depending upon the type of optimizer and learning rate. Neural networks use gradient descent for optimization because the initial point may not be in the neighborhood of the local minimum. Gradient descent has linear convergence as shown in equation 2-20. By contrast, a Newton step shown in equation 6.35 has quadratic convergence, but needs to be in the neighborhood of the optimum. In order to speed up the convergence rate of gradient descent, many algorithms have been proposed. TensorFlow optimizers implement some of the more popular optimization algorithms.

The base class of TensorFlow optimizers is **tf.keras.optimizers. Optimizer**. TensorFlow optimizers are available in module **tf.keras. optimizers**.

$$\Delta\theta = -\alpha\nabla_\theta \tag{2.20}$$

$$\Delta\theta = -\left(\nabla_\theta\nabla_\theta L\right)^{-1}\nabla_\theta \tag{2.21}$$

1. **Adadelta**: This optimizer adapts the learning rate based on the exponentially decaying moving average of gradients and past parameter updates. It is based on research work by Zeiler (2012). Parameter update rule applied by Adadelta is shown in equation 2.22. $\theta$ represents the network's trainable parameters, and $\nabla_\theta L$ is the gradient of loss function $L$ with respect to parameters $\theta$:

$$
\begin{aligned}
g_t &= \nabla_\theta L_t \\
E\left[g_t^2\right] &= \rho E\left[g_{t-1}^2\right] + (1-\rho)g_t^2 \\
E\left[\Delta x_t^2\right] &= \rho E\left[\Delta x_{t-1}^2\right] + (1-\rho)\Delta x_t^2 \\
RMS_{\Delta x} &= \sqrt{E\left[\Delta x_t\right]^2 + \epsilon} \\
RMS_{g_t} &= \sqrt{E\left[g_t^2\right] + \epsilon} \\
\Delta\theta &= -\frac{RMS_{\Delta x}}{RMS_{g_t}} g_t
\end{aligned}
\tag{2.22}
$$

2. **Adam**: This is one of the more widely used optimizers introduced by Kingma and Ba (2014). It uses an adaptive estimate of gradient and gradient square. Square of gradients serves as an approximation to the second-order derivative. Update rule applied by the Adam optimizer is shown in equation 2.23.

In equation 2.23, $\alpha$ is the learning rate, and $\beta_1$ and $\beta_2$ are exponential decay rates for the first- and second-order terms, respectively. The algorithm takes a Newton algorithm like step without explicitly calculating the second derivative (Hessian). To do this, it uses the exponentially decaying moving average of gradient square. $\hat{m}_t$ and $\hat{v}_t$ are the bias-corrected moving average of gradient and gradient square:

$$
\begin{aligned}
g_t &= \nabla_\theta L_t \\
m_t &= \beta_1 m_{t-1} + (1 - \beta_1) g_t \\
v_t &= \beta_2 v_{t-1} + (1 - \beta_2) g_t^2 \\
\hat{m}_t &= \frac{m_t}{1 - \beta_1^t} \\
\hat{v}_t &= \frac{v_t}{1 - \beta_2^t} \\
\Delta\theta &= -\alpha \frac{\hat{m}_t}{\sqrt{\hat{v}_t} + \epsilon}
\end{aligned}
\tag{2.23}
$$

3. **Nadam**: Nadam is an acronym for Nesterov momentum with Adam. As the name suggests, it applies the Nesterov momentum term to the Adam optimizer. It was introduced by Dozat (2016). Equation 2.24 shows the additional Nesterov momentum applied to gradient, $\nabla_\theta L$. After applying momentum, remaining update equations are identical to the Adam update shown in equation 2.23:

$$
g_t = \mu g_{t-1} - \alpha \nabla_\theta L
\tag{2.24}
$$

4. **RMSProp**: RMSProp was proposed by Hinton (2012). It keeps a moving average of square of gradients and uses it as an approximation to Hessian

in performing a Newton-like step. Parameter update equations of the RMSProp optimizer are shown in equation 2.25:

$$g_t = \nabla_\theta L_t$$
$$E\left[g_t^2\right] = \rho E\left[g_{t-1}^2\right] + (1-\rho)g_t^2 \tag{2.25}$$
$$\Delta\theta = -\frac{\alpha}{\sqrt{E\left[g_t^2\right]+\epsilon}}g_t$$

5. **SGD**: This is a general-purpose implementation of the stochastic gradient descent (SGD) algorithm that provides the ability to add momentum. It supports three modes:

- Simple stochastic gradient descent with no momentum as shown in equation 2.20.

- Applying momentum to gradient. Update rule for this version of SGD is shown in equation 2.26, with $\mu$ representing momentum:

$$v_t = \mu v_{t-1} - \alpha \nabla_\theta L$$
$$\Delta\theta = v_t \tag{2.26}$$

- Applying Nesterov momentum to the update equation, as shown in equation 2.27. As before, $\mu$ is momentum:

$$g_t = \nabla_\theta L_t$$
$$v_t = \mu v_{t-1} - \alpha g_t \tag{2.27}$$
$$\Delta\theta = \mu v_t - \alpha g_t$$

Let us compare the performance of optimizers using deep neural networks for classification. The data consists of 13,611 instances of dry bean features such as bean dimensions and shape forms. The objective is to classify the bean into one of seven classes: Seker, Barbunya, Bombay, Cali, Dermason, Horoz, or Sira. The dataset is hosted at the UCI Machine Learning Repository website, "Dry Bean Dataset" (UCI, Dry Bean Dataset). Koklu et al. (2020) created the dataset using computer vision techniques to extract 16 features. All 16 features are numeric, and the result is a categorical variable identifying the bean as one of seven classes.

A neural network model is built and trained for this task. The following observations are noteworthy:

1. Result column "Class" is converted to an integer.

2. Testing/training data partition of 80%/20% is used.

3. Examine the distribution of input features in different classes to make sure all input features are relevant to the classification task. Distributions of four features – perimeter, major axis length, eccentricity, and compactness – are shown as a stacked histogram for the seven classes in Figures 2-23, 2-24, 2-25, and 2-26. As can be seen, all features seem pertinent to the classification task. For example, the "Dermason" class has the lowest perimeter.

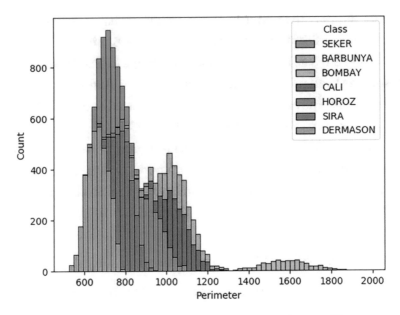

***Figure 2-23.*** *Distribution of Feature Perimeter Across Classes*

4.  Identify any input features with similar distributions
    across classes. This can be done by calculating the
    correlation matrix between normalized features.
    Figures 2-23 and 2-24 show similar distributions.
    Intuitively, perimeter and major axis length are
    likely to be correlated because perimeter is a
    function of major axis length. A joint plot of these
    two features shown in Figure 2-27 confirms the
    hypothesis: notice the elongated shape of the joint
    distribution. Therefore, one of the input features can
    be excluded.

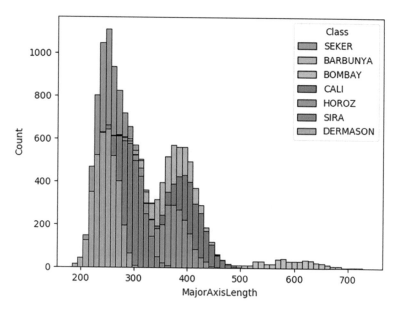

**Figure 2-24.**  *Distribution of Feature Major Axis Length Across Classes*

5.  Numeric feature columns are normalized using training data as $x_{norm} = \dfrac{x - \mu}{2\sigma}$.

6.  This is a multiclass classification problem with seven classes. Sparse categorical cross entropy loss should be used for this problem to conserve space and avoid representing output as a 7-length one-hot vector.

7.  Checkpoint the model for comparable results. Checkpointing assigns initial network weights from a checkpoint file promoting reproducibility of results.

8.  For a multiclass classification problem, a bias
    initializer for the final sigmoid activation layer
    should be set using equation 2.28. It calculates
    the minimum of $\log_e\left(\dfrac{npos_k}{nneg_k}\right)$ over all classes $k$.
    $npos_k$ is the number of positive samples of class $k$
    in the training dataset, and $nneg_k$ is the number of
    remaining samples:

$$\text{bias initializer} = \min_{k \in \text{classes}} \log_e\left(\frac{npos_k}{nneg_k}\right)$$

(2.28)

$nneg_k = N_{data} - npos_k$ where N is number of training data items

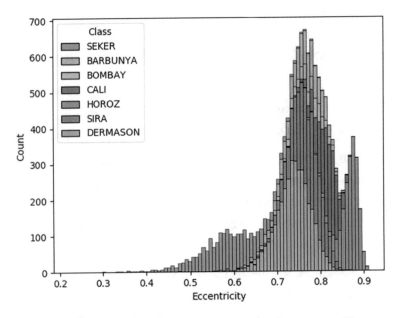

***Figure 2-25.*** *Distribution of Feature Eccentricity Across Classes*

9.  Seven optimizers are considered: Adadelta, Adam,
    Nadam, RMSProp, simple SGD with no momentum,
    SGD with momentum, and SGD with Nesterov
    momentum. The model is fitted over 20 epochs. Loss

function evolution and sparse categorical accuracy are plotted using the seven optimizers in Figures 2-28 and 2-29. As can be seen from the plots, Adam, Nadam, and RMSProp optimizers perform the best. They are followed by the three variants of SGD. Of these three, SGD with no momentum (SGD_simple in the figure) converges the slowest. This shows the benefit of using a momentum term in optimization. Finally, Adadelta performs the worst, indicating that its hyper-parameters need to be tweaked for this problem.

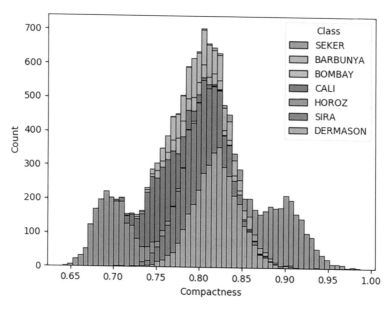

**Figure 2-26.** *Distribution of Feature Compactness Across Classes*

10.  In a multiclass classification problem, the traditional AUC-ROC curve is not directly applicable. Instead, one must look at each class separately and construct an AUC-ROC curve for a binary classification problem of that class vs. the rest.

11.  A confusion matrix for this problem is plotted
     for training and testing data in Figures 2-30 and
     2-31. Since the Adam optimizer gives best-of-class
     performance, these plots are obtained using the
     Adam optimizer.

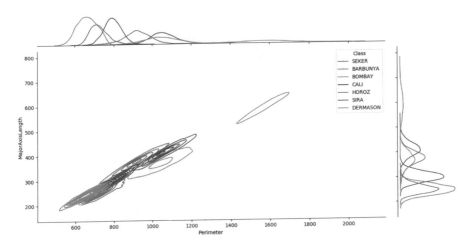

***Figure 2-27.***  *Strongly Correlated Features: Perimeter and Major
Axis Length*

Code for this example is presented in Listing 2-23.

***Listing 2-23.***  Multiclass Classification Problem Comparing
Convergence of Optimizers

```
1    import logging
2    import os
3
4    import matplotlib.pyplot as plt
5    import numpy as np
6    import pandas as pd
7    import seaborn as sns
```

```
8    import tensorflow as tf
9
10   logging.basicConfig(level=logging.DEBUG)
11
12
13   class BeanClassifier(object):
14       """ Classify beans into 1 of 7 classes. Compare
         different optimizers """
15       LOGGER = logging.getLogger(__name__)
16
17       def __init__(self, datadir: str, filename: str =
         "Dry_Bean_Dataset.csv", trainingData: float = 0.8,
18       batchsize: int = 10, epochs: int = 20) -> None:
19           """
20           Initialize
21           :param datadir: Directory name containing
             data file
22           :param filename: dataset file name
23           :param trainingData: Proportion of data to use
             for training
24           :param batchsize: Batch size for gradient descent
25           :param epochs: number of training epochs
26           """
27           df = pd.read_csv(os.path.join(datadir, filename))
28           self.inputDir = datadir
29           self.resultCol = "Class"
30           self.featureCols = list(df.columns)
31           self.featureCols.remove(self.resultCol)
32           self.normalizeCols = {}
33           ntraining = int(trainingData * df.shape[0])
```

```
34      indextrg = np.random.choice(df.shape[0],
        ntraining, replace=False)
35      indextest = np.array([i for i in range(df.
        shape[0]) if i not in set(indextrg)])
36      self.trainDf = df.loc[indextrg, :].reset_
        index(drop=True)
37      self.testDf = df.loc[indextest, :].reset_
        index(drop=True)
38      self.classes = ["Seker", "Barbunya", "Bombay",
        "Cali", "Dermason", "Horoz", "Sira"]
39      self.nClass = len(self.classes)
40      self.classToInt = {k.upper(): i for i, k in
        enumerate(self.classes)}
41      self._normalizeNumericCols(self.trainDf)
42      self.trainDf = self._applyNormalization(self.
        trainDf)
43      self.testDf = self._applyNormalization
        (self.testDf)
44      self.trainDf.loc[:, self.resultCol] = self.
        trainDf.loc[:, self.resultCol].map(self.
        classToInt)
45      self.testDf.loc[:, self.resultCol] = self.testDf.
        loc[:, self.resultCol].map(self.classToInt)
46      self.metrics = [tf.keras.metrics.
        SparseCategoricalAccuracy()]
47      self.batchSize = batchsize
48      self.nEpoch = epochs
49      self.optimizers = [tf.keras.optimizers.
        Adadelta(),
50      tf.keras.optimizers.Adam(),
51      tf.keras.optimizers.Nadam(),
```

```
52          tf.keras.optimizers.RMSprop(),
53          tf.keras.optimizers.SGD(name="SGD_simple"),
54          tf.keras.optimizers.SGD(momentum=0.1,
            name="SGD_mom"),
55          tf.keras.optimizers.SGD(momentum=0.1,
            nesterov=True, name="SGD_nest_mom")]
56
57      def _normalizeNumericCols(self, trainingDf:
        pd.DataFrame) -> None:
58          """
59          Calclate normalizing params for numeric columns
60          :param trainingDf:
61          :return: None
62          """
63          for col in self.featureCols:
64              mean = trainingDf.loc[:, col].mean()
65              sd = trainingDf.loc[:, col].std()
66              self.normalizeCols[col] = (mean, 2*sd)
67
68      def _applyNormalization(self, df: pd.DataFrame) ->
        pd.DataFrame:
69          """
70          Apply normalization as col = (x-mean)/(2*sd)
71          :param df:
72          :return: df
73          """
74          for col in self.featureCols:
75              mean, sd2 = self.normalizeCols[col]
76              df.loc[:, col] = (df.loc[:, col].values -
                mean) / sd2
77          return df
```

```
78
79      def _getInitializer(self) -> tf.keras.initializers.
        Constant:
80          """
81          Get initializer of final layer
82          :return:
83          """
84          minval = 0
85          for i in range(self.nClass):
86              npos = (self.trainDf.loc[:, self.resultCol]
                == i).sum()
87              nneg = self.trainDf.shape[0] - npos
88              if nneg != 0:
89                  initval = np.log(npos / float(nneg))
90                  if (minval == 0) or (minval > initval):
91                      minval = initval
92          return tf.keras.initializers.Constant(minval)
93
94      def model(self, optimizer: tf.keras.optimizers.
        Optimizer) -> tf.keras.Model:
95          """
96          Create a neural network model for classification
            and initialize weights from a
97          saved checkpoint
98          :param optimizer: Optimizer to use in the model
99          :return: Neural network model
100         """
101         nnet = tf.keras.models.Sequential()
102         initializer = self._getInitializer()
103         nnet.add(tf.keras.layers.Dense(10,
            activation="relu", input_shape=(len(self.
            featureCols),)))
```

```
104         nnet.add(tf.keras.layers.Dense(20,
            activation="relu"))
105         nnet.add(tf.keras.layers.Dense(self.
            nClass, activation="sigmoid", bias_
            initializer=initializer))
106         nnet.compile(optimizer=optimizer,
107         loss=tf.keras.losses.
            SparseCategoricalCrossentropy(),
108         metrics=self.metrics)
109         self.checkpointModel(nnet)
110         return nnet
111
112     def checkpointModel(self, nnet):
113         checkpointFile = os.path.join(self.inputDir,
            "checkpoint_dbean_wt")
114         if not os.path.exists(checkpointFile):
115             nnet.predict(np.ones((20, len(self.
                featureCols)), dtype=np.float32))
116             tf.keras.models.save_model(nnet,
                checkpointFile, overwrite=False)
117         else:
118             nnet = tf.keras.models.load_
                model(checkpointFile)
119         return nnet
120
121     def optimizerConvergence(self):
122         histDict = {}
123         for opt in self.optimizers:
124             nnet = self.model(opt)
125             nnet, history = self.trainModel(nnet)
126             histDict[opt._name] = history
```

```
127              if opt._name == "Adam":
128                  self.testModel(nnet)
129          for metric in self.metrics:
130              self.plotConvergenceHistory(histDict,
                 metric._name)
131          self.plotConvergenceHistory(histDict, "loss")
132
133      def plotConvergenceHistory(self, histDict,
         metricName):
134          for name, history in histDict.items():
135              plt.plot(history.epoch, history.
                 history[metricName], label=name)
136
137          plt.xlabel("Epoch")
138          plt.ylabel(metricName)
139          plt.grid(True)
140          plt.legend()
141          plt.show()
142
143      def plotConfusionMatrix(self, labels: np.ndarray,
         predictions: np.ndarray) -> None:
144          predictedLabels = np.argmax(predictions, axis=1)
145          fig, ax = plt.subplots()
146          cm = np.zeros((self.nClass, self.nClass),
                 dtype=np.int32)
147          for i in range(labels.shape[0]):
148              cm[labels[i], predictedLabels[i]] += 1
149          sns.heatmap(cm, annot=True, fmt="d",
                 linewidths=0.25, ax=ax)
150          ax.set_xticks(range(1+self.nClass))
151          ax.set_yticks(range(1+self.nClass))
```

```
152            ax.set_xticklabels(["0"] + self.classes)
153            ax.set_yticklabels(["0"] + self.classes)
154            ax.set_ylabel('Actual')
155            ax.set_xlabel('Predicted')
156            plt.show()
157
158        def testModel(self, nnet):
159            for df in [self.trainDf, self.testDf]:
160                features = df.loc[:, self.featureCols].values
161                actClass = df.loc[:, self.resultCol].values
162                predictClass = nnet.predict(features)
163                self.plotConfusionMatrix(actClass,
                   predictClass)
164
165        def trainModel(self, nnet):
166            trainFeatures = self.trainDf.loc[:, self.
               featureCols].values
167            trainClass = self.trainDf.loc[:, self.
               resultCol].values
168            history = nnet.fit(trainFeatures, trainClass,
               batch_size=self.batchSize, epochs=self.nEpoch)
169            return nnet, history
170
171    if __name__ == "__main__":
172        bclassify = BeanClassifier(r"C:\prog\cygwin\home\
           samit_000\RLPy\data\book\DryBeanDataset")
173        bclassify.optimizerConvergence()
```

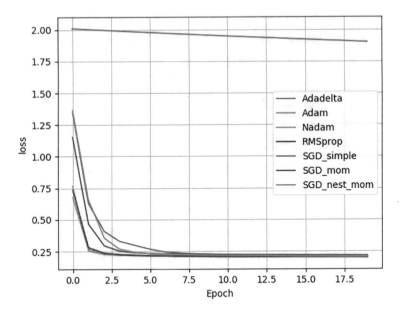

**Figure 2-28.** *Evolution of Loss for Seven Optimizers in Multiclass Classification*

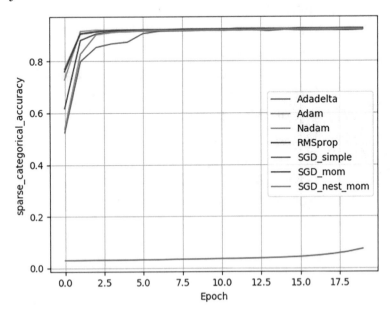

**Figure 2-29.** *Evolution of Sparse Categorical Accuracy for Seven Optimizers in Multiclass Classification*

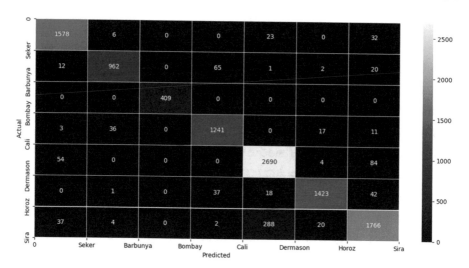

***Figure 2-30.*** *Confusion Matrix for Training Data Using the Adam Optimizer*

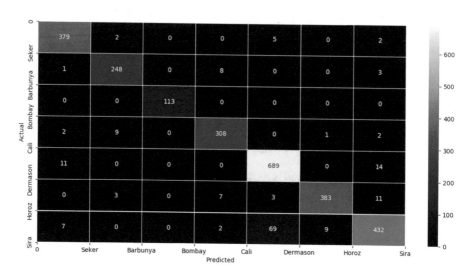

***Figure 2-31.*** *Confusion Matrix for Testing Data Using the Adam Optimizer*

# 2.10 Regularizers

Regularization prevents overfitting of model parameters to the training dataset. Regularization adds another component to the loss function so that optimization does not focus solely on reducing the difference between actual outputs and predicted outputs. By reducing overfitting, regularization may improve model performance in the testing dataset.

In TensorFlow, regularization is applied to layers because a model's trainable parameters are stored in layers. There are two kinds of regularization: weight regularization and activity regularization. Because TensorFlow layers keep bias weights and neuron connection weights separately, weight regularization is subdivided into two types: kernel regularization and bias regularization. These three regularizations are described in the following and illustrated with an example later in this section.

1.  **Kernel regularization**: Applies a regularization penalty to layer weights excluding the bias weight. This is specified using the **kernel_regularizer** argument to a layer's constructor.

2.  **Bias regularization**: Applies a regularization penalty to bias weights using the **bias_regularizer** argument to a layer's constructor.

3.  **Activity regularization**: A regularization penalty is applied to the layer's output after normalizing the term by batch size. This can be specified using the **activity_regularizer** argument to a layer's constructor.

TensorFlow has the following regularizers available for use, all defined in module **tf.keras.regularizers**:

1.  **tf.keras.regularizers.L1**: Applies L1 regularization as $\sum_i |w_i|$ over weights $w_i$. The constructor takes a regularization penalty weight.

2. **tf.keras.regularizers.L2**: Applies L2 regularization as $\sum_i w_i^2$ over weights $w_i$. The constructor takes a regularization penalty weight.

3. **tf.keras.regularizers.L1L2**: Applies a weighted sum of L1 and L2 regularizations with the constructor specifying the weights.

In order to define a custom regularization, derive a class from base class **tf.keras.regularizer.Regularizer** and provide an implementation of the __call__ method that takes weights as an argument. An example is shown in Listing 2-24 that defines a custom regularizer as a weighted sum of L4 and L1 regularizations. There is one data point in input with five features. The layer has two neurons (units). With a multiplier of 0.01 for both L1 and L2 terms, this should give a loss of $0.01 * 10 + 0.01 * 10$ or 0.2, as is seen in the output. Method **get_config** is used in serialization/deserialization of a regularizer.

***Listing 2-24.*** Custom L1L4 Regularizer

```
1    import tensorflow as tf
2
3
4    class L1L4Regularizer(tf.keras.regularizers.Regularizer):
5        def __init__(self, l1=0.01, l4=0.01):
6            self.l1 = l1
7            self.l4 = l4
8
9        def __call__(self, weights):
10           sq = tf.math.square(weights)
11           return self.l1 * tf.math.reduce_sum(tf.math.
             abs(weights)) + \
12           self.l4 * tf.math.reduce_sum(tf.math.square(sq))
```

```
13
14        def get_config(self):
15            return {"l1": self.l1, "l4": self.l4}
16
17    layer = tf.keras.layers.Dense(2, input_shape=(5,), kernel_
      regularizer=L1L4Regularizer(),
18    kernel_initializer="ones")
19    input = tf.ones(shape=(1, 5))
20    output = layer(input)
21    print(layer.losses)
22
23    [<tf.Tensor: id=121, shape=(), dtype=float32,
      numpy=0.19999999>]
```

In the following example, let us look at the "Auto MPG Dataset" available in the UCI Machine Learning Repository (2022). The dataset was used by Quinlan (1993). It is a regression problem of predicting an automobile's MPG (miles-per-gallon) fuel consumption. It has eight input features and 398 data points. A neural network model is built for the regression task as follows:

1.  Of the input features, three are categorical: cylinders, model year, and origin. Since they are integers, they do not need to be normalized.

2.  Input feature **car name** is a string. This feature is particularly prone to data errors and misformatting. For example, automobile manufacturer occurs as **mercedes benz** and **mercedes-benz** for two different cars. A careful data processing module is needed to process this feature. In this model, a simplistic approach is adopted: only the first two words of the string are considered and converted to

lowercase. This two-word string is then processed as a categorical variable. The reasoning behind this approach is that car name consists of car manufacturer followed by model and other optional qualifiers, like **toyota corolla 1200**. The model uses **toyota corolla** as a categorical feature.

3. Feature **horsepower** has missing values that are replaced with 0.

4. A training-testing split of 80%-20% is used. This gives 314 training data points.

5. Training data consists of relatively few points – 314. Due to this, the model should have as few parameters as possible. The features are plotted against the result (MPG) to ascertain they are all relevant for the regression task.

6. Correlation of numeric features **displacement, horsepower, weight**, and **acceleration** against output **mpg** is shown in Figure 2-32. The plot shows that all numeric columns are relevant for regression; there is no feature with a small absolute correlation coefficient. If a feature with a small absolute correlation against output is found, it should be dropped to keep the model parsimonious.

7. Numeric features are normalized using $\dfrac{x - \mu}{2\sigma}$ where $\mu$ and $\sigma$ are the mean and standard deviation of feature $x$ in training data, as before.

8. A correlation matrix of numeric input features is calculated. As seen from the heatmap in Figure 2-33, no correlation coefficient between input features

is too high – for example, about 0.8. If two input features are highly correlated, consider dropping one of them.

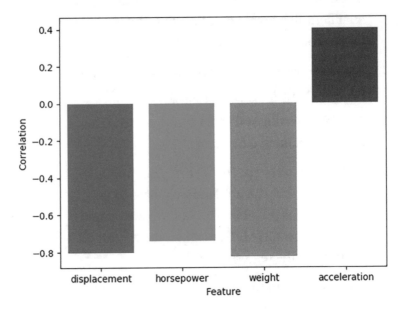

***Figure 2-32.*** *Correlation of Numeric Input Features Against the Output Variable*

9.  A histogram of the output variable **mpg** is plotted as a histogram for categorical feature values as shown in Figures 2-34, 2-35, 2-36, and 2-37. The first three categorical variables seem relevant for regression. For example, **mpg** is lower for a higher number of cylinders (Figure 2-34) . Similarly, **mpg** is higher for later years (Figure 2-35). However, there seems to be no clear relation between **car name** and **mpg** (Figure 2-37). Therefore, categorical variable **car name** is dropped.

10.  A neural network model is built for three cases: no regularization, L1 regularization, and L2 regularization applied to neuron weights in all layers (**kernel_regularizer**). Loss evolution and metric (mean absolute error) evolution is plotted for test data for the three cases. As can be seen from Figures 2-38 and 2-39, there is a small benefit to including regularization – though not much for this regression problem.

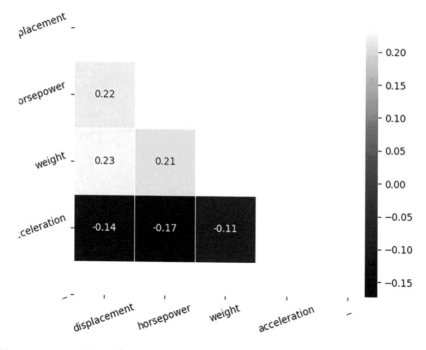

**Figure 2-33.**  *Correlation Matrix of Numeric Input Features – No Highly Correlated Features*

11. Predicted vs. actual results for training and testing
datasets for the three runs (no regularizer, L1
regularizer, and L2 regularizer) are shown in
Figures 2-40 to 2-45. The model performs a decent
task of predicting the *mpg* of a car.

The code for this example is shown in Listing 2-25.

***Listing 2-25.*** Regression Problem of Predicting Automobile MPG
Fuel Consumption and Comparing Regularizers

```
1    import numpy as np
2    import pandas as pd
3    import tensorflow as tf
4    import logging
5    import os
6    import matplotlib.pyplot as plt
7    import seaborn as sns
8    import copy
9    from typing import List
10
11   logging.basicConfig(level=logging.DEBUG)
12
13
14   class AutoMPG(object):
15       "Neural network model for regression. Predict
         automobile MPG"
16       LOGGER = logging.getLogger("AutoMPG")
17
18       def __init__(self, datadir: str, filename: str =
         "auto-mpg.data", trainingData: float = 0.8,
19                    batchSize: int = 10, epochs: int = 20):
```

```
20          self.columns = ["mpg", "cylinders",
            "displacement", "horsepower", "weight",
            "acceleration", "model year",
21                          "origin", "car name"]
22          df = pd.read_csv(os.path.join(datadir, filename),
            header=None, names=self.columns, sep="\s+")
23          self.inputDir = datadir
24          self.batchSize = batchSize
25          self.nEpoch = epochs
26          df = self._preprocessData(df)
27          self.resultCol = "mpg"
28          self.categoricalCols = ["cylinders", "model
            year", "origin", "car name"]
29          exclude = set(self.categoricalCols)
30          self.numericCols = [c for c in df.columns if c
            not in exclude]
31          self.featureCols = self.numericCols + self.
            categoricalCols
32          self.featureCols.remove(self.resultCol)
33          self.normalizeCols = {}
34          self.categoricalMap = {}
35          ntrain = int(trainingData * df.shape[0])
36          trainIndex = np.random.choice(df.shape[0],
            ntrain, replace=False)
37          testIndex = np.array([i for i in range(df.
            shape[0]) if i not in set(trainIndex)])
38          self.trainDf = df.loc[trainIndex, :].reset_
            index(drop=True)
39          self.testDf = df.loc[testIndex, :].reset_
            index(drop=True)
40          self._normalizeNumericCols(self.trainDf)
```

```
41          self.trainDf = self._applyNormalization(self.
            trainDf)
42          self.testDf = self._applyNormalization
            (self.testDf)
43          corr1 = self._correlWithOutput(self.trainDf)
44          cov = self._correlWithinInputs(self.trainDf)
45          self._processCategoricalCols(df)
46          self.trainDf = self._applyCategoricalMapping
            (self.trainDf)
47          self.testDf = self._applyCategoricalMapping
            (self.testDf)
48          self.metrics = [tf.keras.metrics.
            MeanAbsoluteError()]
49          self.featureCols.remove("car name")
50          self.categoricalCols.remove("car name")
51
52      def _processCategoricalCols(self, df: pd.DataFrame)
        -> None:
53          """
54          Process categorical columns by creating a mapping
55          :param df: training dataframe
56          :rtype: None
57          """
58          for col in self.categoricalCols:
59              unique = np.sort(df.loc[:, col].unique())
60              self.categoricalMap[col] = {u:i for i,u in
                enumerate(unique)}
61
62      def _correlWithOutput(self, df: pd.DataFrame) ->
        np.ndarray:
63          output = df.loc[:, self.resultCol].values
```

```
64              mu = output.mean()
65              sd = output.std()
66              x = (output - mu)/sd
67              ncols = copy.copy(self.numericCols)
68              ncols.remove(self.resultCol)
69              correl = np.zeros(len(ncols), dtype=np.float32)
70              for i, col in enumerate(ncols):
71                  x1 = df.loc[:, col].values
72                  correl[i] = np.sum(2 * x * x1)/df.shape[0]
73              plotdf = pd.DataFrame({"Feature": ncols,
                "Correlation": correl})
74              sns.barplot(x="Feature", y="Correlation",
                data=plotdf)
75              plt.show()
76
77              mean, sd = self.normalizeCols[self.resultCol]
78              mpg = df.loc[:, self.resultCol].values *
                sd + mean
79              for col in self.categoricalCols:
80                  sns.histplot(data=df, x=mpg, hue=col)
81                  plt.show()
82              return correl
83
84          def _correlWithinInputs(self, df: pd.DataFrame) ->
            np.ndarray:
85              ncols = copy.copy(self.numericCols)
86              ncols.remove(self.resultCol)
87              cov = np.cov(df.loc[:, ncols].values.T)
88              mask = np.triu(np.ones_like(cov, dtype=bool))
89              fig, ax = plt.subplots()
```

```
90          sns.heatmap(cov, mask=mask, annot=True,
            linewidths=0.25, ax=ax)
91          ax.set_xticks(0.5 + np.arange(cov.shape[0]+1))
92          ax.set_yticks(0.5 + np.arange(cov.shape[0]+1))
93          ax.set_xticklabels(ncols + ["_"], rotation=20)
94          ax.set_yticklabels(ncols + ["_"], rotation=20)
95          plt.show()
96          return cov
97
98      def _preprocessData(self, df: pd.DataFrame) ->
        pd.DataFrame:
99          df.loc[:, "horsepower"] = df.loc[:,
            "horsepower"].replace("?", 0).astype(np.float32)
100         func = lambda x: x.lower().split(" ", 3)[0]
101         df.loc[:, "car name"] = df.loc[:, "car name"].
            map(func)
102         return df
103
104     def _normalizeNumericCols(self, trainingDf:
        pd.DataFrame) -> None:
105         """
106         Calclate normalizing params for numeric columns
107         :param trainingDf:
108         :return: None
109         """
110         for col in self.numericCols:
111             mean = trainingDf.loc[:, col].mean()
112             sd = trainingDf.loc[:, col].std()
113             self.normalizeCols[col] = (mean, 2*sd)
114
```

```
115    def _applyNormalization(self, df: pd.DataFrame) ->
       pd.DataFrame:
116        """
117        Apply normalization as col = (x-mean)/(2*sd)
118        :param df:
119        :return: df
120        """
121        for col in self.numericCols:
122            mean, sd2 = self.normalizeCols[col]
123            df.loc[:, col] = (df.loc[:, col].values -
               mean) / sd2
124        return df
125
126    def _applyCategoricalMapping(self, df: pd.DataFrame)
       -> pd.DataFrame:
127        """
128        Apply mapping to convert categorical columns to
           integers
129        :rtype: pd.DataFrame with mapped
           categorical columns
130        """
131        for col in self.categoricalCols:
132            df.loc[:, col] = df.loc[:, col].map(self.
               categoricalMap[col])
133        return df
134
135    def testRegularizers(self, regularizers: List[tf.
       keras.regularizers.Regularizer], names: List[str])
       -> None:
136        histDict = {}
```

```
137         for regularizer, name in
            zip(regularizers, names):
138             self.nnet = self.
                model(regularizer=regularizer)
139             history = self.trainModel()
140             histDict[name] = history
141             self.testModel(name)
142
143         for metric in self.metrics:
144             self.plotConvergenceHistory(histDict,
                metric._name)
145         self.plotConvergenceHistory(histDict, "loss")
146
147     def model(self, regularizer: tf.keras.regularizers.
        Regularizer = None) -> tf.keras.Model:
148         nfeature = len(self.featureCols)
149         nnet = tf.keras.models.Sequential()
150         nnet.add(tf.keras.layers.Dense(12,
            activation="sigmoid", input_shape=(nfeature,),
            kernel_regularizer=regularizer))
151         nnet.add(tf.keras.layers.
            Dense(3, activation="relu", kernel_
            regularizer=regularizer))
152         nnet.add(tf.keras.layers.Dense(1, kernel_
            regularizer=regularizer))
153         nnet.compile(optimizer=tf.keras.optimizers.
            Adam(learning_rate=0.002),
154                     loss=tf.keras.losses.
                        MeanSquaredError(),
155                     metrics=self.metrics)
156         nnet = self.checkpointModel(nnet)
```

```
157            return nnet
158
159        def checkpointModel(self, nnet):
160            checkpointFile = os.path.join(self.inputDir,
                   "checkpoint_autompg_wt")
161            if not os.path.exists(checkpointFile):
162                nfeature = len(self.featureCols)
163                nnet.predict(np.ones((20, nfeature),
                       dtype=np.float32))
164                tf.keras.models.save_model(nnet,
                       checkpointFile, overwrite=False)
165            else:
166                nnet = tf.keras.models.load_model
                       (checkpointFile)
167            return nnet
168
169        def trainModel(self, trainDf: pd.DataFrame = None) ->
               tf.keras.callbacks.History:
170            if trainDf is None:
171                trainDf = self.trainDf
172            X = trainDf.loc[:, self.featureCols].values
173            y = trainDf.loc[:, self.resultCol].values
174            history = self.nnet.fit(X, y, batch_size=self.
                   batchSize, epochs=self.nEpoch)
175            return history
176
177        def testModel(self, title: str) -> None:
178            loss = tf.keras.losses.MeanSquaredError()
179            for df in [self.trainDf, self.testDf]:
180                features = df.loc[:, self.featureCols].values
181                actVals = df.loc[:, self.resultCol].values
```

```
182            predictVals = self.nnet.predict(features)
183            lossval = loss(actVals[:, np.newaxis],
               predictVals)
184            self.LOGGER.info("Loss for regularizer
               %s, number of data points %d: %f", title,
               df.shape[0], lossval)
185            self.plotActualVsPredicted(actVals,
               predictVals.squeeze(), title=title)
186
187    def plotActualVsPredicted(self, actualVals:
       np.ndarray, predictedVals: np.ndarray, title: str =
       None) -> None:
188        mean, sd = self.normalizeCols[self.resultCol]
189        y = actualVals * sd + mean
190        x = predictedVals * sd + mean
191        plt.scatter(x, y, c="red")
192        p1 = max(max(x), max(y))
193        p2 = min(min(x), min(y))
194        plt.plot([p1, p2], [p1, p2], 'b-')
195        plt.xlabel("Predicted Values")
196        plt.ylabel("Actual Values")
197        if title:
198            plt.title(title)
199        plt.show()
200
201    def plotConvergenceHistory(self, histDict: dict,
       metricName: str) -> None:
202        for name, history in histDict.items():
203            plt.plot(history.epoch, history.
               history[metricName], label=name)
204        plt.xlabel("Epoch")
```

```
205            plt.ylabel(metricName)
206            plt.grid(True)
207            plt.legend()
208            plt.show()
209
210
211    if __name__ == "__main__":
212        mpg = AutoMPG(r"C:\prog\cygwin\home\samit_000\RLPy\
               data\book", batchSize=1)
213        regularizers = [None, tf.keras.regularizers.
               L1L2(l1=0.1, l2=0), tf.keras.regularizers.L1L2(l1=0,
               l2=0.1)]
214        names = ["None", "L1", "L2"]
215        mpg.testRegularizers(regularizers, names)
```

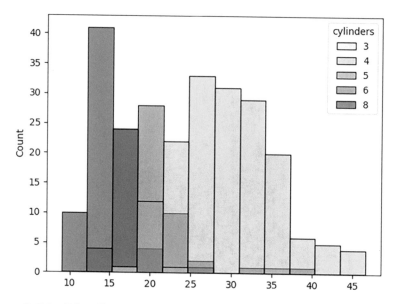

**Figure 2-34.** *Distribution of Output MPG Against Cylinders*

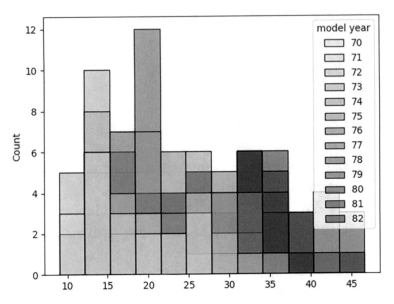

***Figure 2-35.*** *Distribution of Output MPG Against Year*

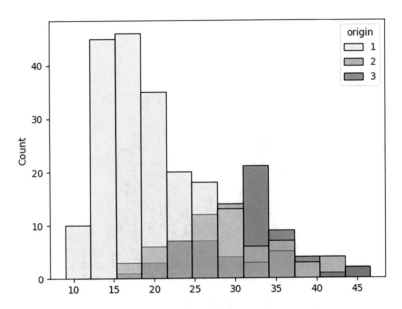

***Figure 2-36.*** *Distribution of Output MPG Against Origin*

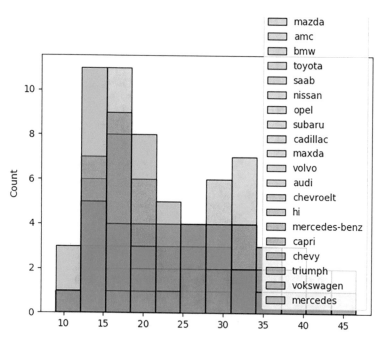

***Figure 2-37.***  *Distribution of Output MPG Against Car Name*

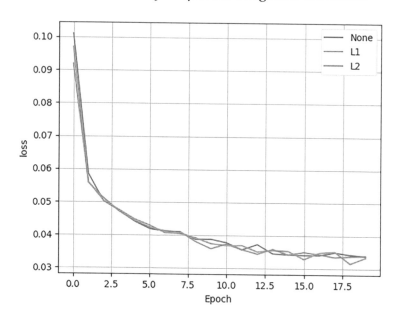

***Figure 2-38.***  *Evolution of Mean Square Error Loss with Epochs*

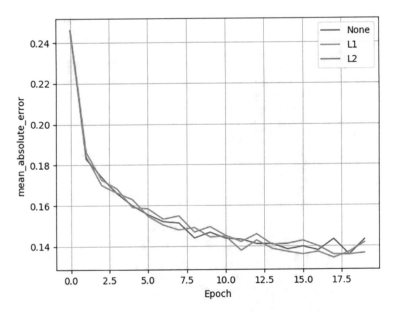

***Figure 2-39.*** *Evolution of Mean Absolute Error Metric with Epochs*

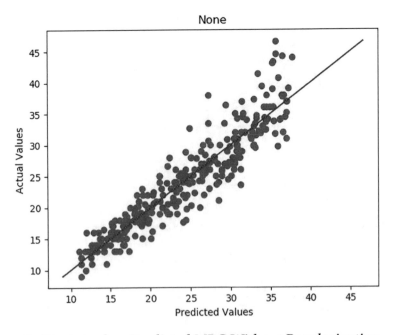

***Figure 2-40.*** *Actual vs. Predicted MPG Without Regularization,*
*Training Set*

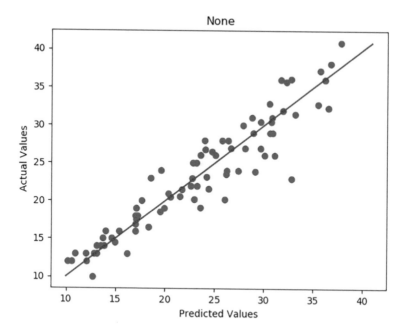

***Figure 2-41.*** *Actual vs. Predicted MPG Without Regularization, Testing Set*

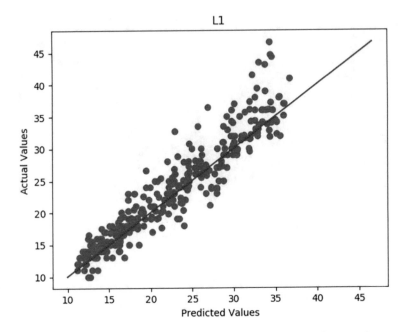

**Figure 2-42.**  *Actual vs. Predicted MPG with L1 Regularization, Training Set*

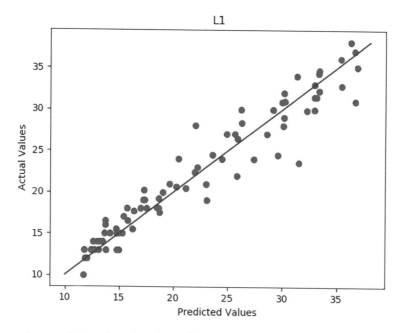

***Figure 2-43.*** *Actual vs. Predicted MPG with L1 Regularization, Testing Set*

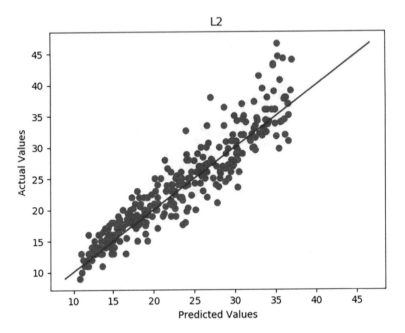

**Figure 2-44.**  *Actual vs. Predicted MPG with L2 Regularization, Training Set*

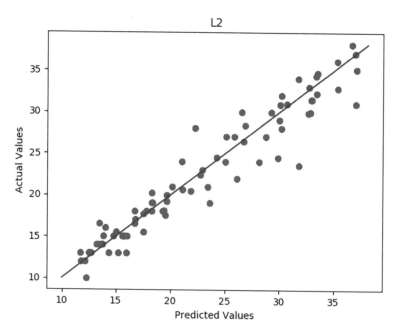

**Figure 2-45.** *Actual vs. Predicted MPG with L2 Regularization, Testing Set*

# 2.11 TensorBoard

TensorBoard is a visualization tool for working with TensorFlow models, providing a comprehensive toolkit for profiling, visualizing metrics, and analyzing layers of a deep neural network. It is loaded as an extension into Jupyter Notebook using the command **%load_ext tensorboard**. Once loaded, the TensorFlow neural network model is compiled. Before calling fit, a callback is provided as an argument to method **fit**. The callback will record training history in a log directory. After model training is complete, TensorBoard is launched from Jupyter Notebook using the command **%tensorboard -logdir logs/fit**. TensorBoard can also be launched as a standalone utility outside Jupyter Notebook using the command **tensorboard -logdir logs/fit**.

Using Jupyter Notebook, the steps for using TensorBoard are summarized in the following and illustrated in Listing 2-26:

1. Load the TensorBoard extension into Jupyter Notebook.

2. Create a log directory for writing logs.

3. Create a TensorBoard callback and specify the log directory.

4. Pass the callback as an argument to the **fit** method.

5. Once training is complete, launch the TensorBoard user interface using **%tensorboard –logdir log_dir**. Specify the log directory.

***Listing 2-26.*** Using TensorBoard

```
1    import tensorflow as tf
2    import numpy as np
3    import datetime
```

```
4    import shutil
5
6    # load tensorboard extension
7    %load_ext tensorboard
8
9    # specify base logs dir
10   base_log_dir = "logs\\fit\\"
11
12   # clear previous logs
13   try:
14       shutil.rmtree(base_log_dir)
15   except OSError as e:
16       pass
17
18   # create some data
19   nfeature = 10
20   nsample = 100
21   nsampletest = 20
22   X = 1 + np.random.random((nsample, nfeature))
23   y = 2*X.sum(axis=1) + 4
24
25   Xtest = 1 + np.random.random((nsampletest, nfeature))
26   ytest = 2*Xtest.sum(axis=1) + 4
27
28   nnet = tf.keras.models.Sequential()
29   nnet.add(tf.keras.layers.Dense(4, input_
     shape=(nfeature,)))
30   nnet.add(tf.keras.layers.Dense(10, activation="relu"))
31   nnet.add(tf.keras.layers.Dropout(0.2))
32   nnet.add(tf.keras.layers.Dense(1))
33
```

```
34   nnet.compile(optimizer="adam", loss="MSE", metrics=[tf.
     keras.metrics.MeanAbsoluteError()])
35
36   # specify log directory
37   log_dir = base_log_dir + datetime.datetime.now().
     strftime("run%Y%m%d_%H%M%S")
38
39   # create TensorBoard callback
40   tb_callback = tf.keras.callbacks.TensorBoard(log_dir=log_
     dir, histogram_freq=1)
41
42   # provide the callback to fit method
43   nnet.fit(X, y, epochs=10, callbacks=[tb_callback])
```

## 2.12  Dataset Manipulation

TensorFlow provides APIs for creating datasets, creating batches,
processing input features, applying mapping, and shuffling items in the
module **tf.keras.Dataset**. This module can be used to create an input
pipeline for a neural network model. It can work with large datasets that do
not fit in memory all at once by streaming the data as needed.

A dataset can be created using different data sources. A few commonly
used data sources are shown in the following:

1.  **From Python list or numpy array objects** using
    **from_tensor_slices** as shown in Listing 2-27.

2.  **From file(s)**, a dataset can be created using **tf.data.
    TextLineDataset**.

3.  **Range** using **tf.data.Dataset.range**.

***Listing 2-27.*** Creating a Dataset

```
1   import numpy as np
2   import tensorflow as tf
3   import pandas as pd
4
5   def create_dataset():
6       # from list and numpy array
7       lst = [4] * 4
8       ar = np.ones(1, dtype=np.int32) * 4
9       dset1 = tf.data.Dataset.from_tensor_slices(lst)
10      dset2 = tf.data.Dataset.from_tensor_slices(ar)
11      for e1, e2 in zip(dset1, dset2):
12      assert e1 == e2
13
14      # from csv file
15      df = pd.DataFrame({"a": [1, 2, 3], "b": ["r1",
        "r2", "r3"]})
16      filename = r"C:\prog\cygwin\home\samit_000\RLPy\data\
        book\test.csv"
17      df.to_csv(filename, index=False)
18      dataset = tf.data.TextLineDataset([filename])
19      for row in dataset:
20          print(row)
21
22  if __name__ == "__main__":
23      create_dataset()
```

Useful methods of the **Dataset** class are described in the following:

1. **from_tensor_slices**: Creates a dataset from an iterable along the first dimension. For example, **from_tensor_slices(arr)** where **arr** is a numpy array of shape (2, 4, 5) will create a dataset of two tensors of shape (4, 5) each. This method accepts multiple iterable objects.

2. **from_tensors**: Creates a dataset from provided tensors.

3. **map**: Apply a mapping function to each tensor within the dataset.

4. **batch**: Creates a set of batches each with a specified number of elements from the dataset. The last batch can have fewer elements if the dataset size is not divisible by batch size. This is shown in Listing 2-28.

*Listing 2-28.* Batching

```
1    import tensorflow as tf
2    dset = tf.data.Dataset.range(5)
3    batches = dset.batch(2)
4    for batch in batches:
5        print(batch)
6
7    tf.Tensor([0 1], shape=(2,), dtype=int64)
8    tf.Tensor([2 3], shape=(2,), dtype=int64)
9    tf.Tensor([4], shape=(1,), dtype=int64)
```

5. **concatenate**: Concatenates a dataset with another dataset.

6. **shard**: Gives a dataset containing a subset of elements such that i mod N = 0 where **i** is the element's index and **N** is the argument to the shard method. An example is shown in Listing 2-29.

***Listing 2-29.*** Sharding a Dataset

```
1   import tensorflow as tf
2   dset2 = tf.data.Dataset.range(10)
3   shard = dset2.shard(num_shards=3, index=0)
4   for element in shard:
5       print(element)
6
7   tf.Tensor(0, shape=(), dtype=int64)
8   tf.Tensor(3, shape=(), dtype=int64)
9   tf.Tensor(6, shape=(), dtype=int64)
10  tf.Tensor(9, shape=(), dtype=int64)
```

7. **shuffle**: Shuffles the elements of a dataset by selecting **buffer_size** number of elements randomly from the dataset, randomly drawing elements from the buffer, and replacing the drawn elements with new elements from the dataset. To ensure that the dataset is shuffled in a uniform random fashion, that is, with the probability of any element being selected as $\frac{1}{N}$, buffer size must be greater than or equal to the number of elements in dataset, $N$.

8. **repeat**: Concatenate a specified number of copies of a dataset.

# 2.13 Gradient Tape

Gradient tape is used in TensorFlow to perform automatic differentiation of the loss function with respect to tensors. In backpropagation, the loss function is differentiated with respect to output layer weights first, followed by the next layer and so on. This is because the loss function is defined in terms of network output and actual output. Network output is produced by the output layer. Gradients are computed from the output layer and propagated backward toward the input layer using the chain rule. Gradient tape is instrumental in backpropagating the gradients automatically across the layers of a neural network.

Gradient tape remembers operations on tensors during forward pass, as input is fed to the input layer and gets propagated through the network layers, producing a network output. During backpropagation, the loss function is calculated and differentiated with respect to trainable layer weights that are stored as **tf.Variable**.

Most neural network problems do not require a programmer to use GradientTape explicitly; it is implicitly used inside the neural network's **fit** method. However, there are cases where one needs to write a custom loss function that depends on other neural networks. In such cases, GradientTape must be used.

By default, GradientTape watches all trainable **tf.Variable** objects. The neural network's trainable weights, being the constituent layers' trainable weights, are objects of type **tf.Variable** and are watched by GradientTape. Any mathematical calculation involving watched variables is recorded by the tape for subsequent differentiation. Mathematical calculations must use TensorFlow functions whose derivatives are known. To disable the default behavior of watching **tf.Variable** objects, the **watch_accessed_variable** argument of GradientTape's constructor can be set to **False**. To watch a tensor, use the **watch** method of GradientTape. Using GradientTape to calculate gradients requires a programmer to be cognizant of the following features:

1. To calculate gradients, use GradientTape's **gradient** method. This method returns gradients of the same shape as variables with respect to which gradient is calculated. This is illustrated in Listing 2-30.

***Listing 2-30.*** Shape of Gradient

```
1   import tensorflow as tf
2   X = tf.constant(tf.random.normal((5, 4)))
3   W = tf.Variable(tf.ones((4, 6), dtype=tf.float32))
    # watched by default
4   b = tf.constant(tf.ones(6, dtype=tf.float32))
    # not watched by default
5   with tf.GradientTape() as tape:
6       tape.watch(b)
7       y = tf.matmul(X, W) + b
8   vars = [W, b]
9   grads = tape.gradient(y, vars)
10  for i, grad in enumerate(grads):
11      print(f"Variable shape: {vars[i].shape},
            gradient shape: {grad.shape}")
12
13  Variable shape: (4, 6), gradient shape: (4, 6)
14  Variable shape: (6,), gradient shape: (6,)
```

2. Tensors are not watched by default and must be explicitly added to the variables watched by GradientTape, as seen in Listing 2-30.

3. Targets that do not have functional dependence on a variable will give None as gradient.

4. To update a **tf.Variable**, use the **assign** method of **tf.Variable**.

5. Perform calculations using TensorFlow math library functions and not external library functions (such as numpy).

6. Gradients must be computed with respect to floating-point variables that are watched by GradientTape. Using string or integer variable types will give a **None** gradient.

Let us apply these principles on a practical classification problem. **tf.keras.datasets.fashion_mnist** is a dataset of 70,000 clothing item images belonging to a set of ten classes: top, trouser, pullover, dress, coat, sandal, shirt, sneaker, bag, and boot. Each image has size of 28 by 28 pixels. Of the 70,000 images, 60,000 are in the training dataset, and 10,000 are in the testing dataset. A few salient features of the model are noted in the following:

1. Images are converted into a decimal format by dividing by 255, which is the maximum pixel value.

2. A simple neural network model is trained on this dataset using gradient tape. The model first flattens the image from 28 by 28 pixels to an array of 784 input features.

3. The model uses the sparse categorical cross entropy loss function to conserve space. This loss allows actual output to be specified as an integer class label instead of a one-hot vector of size 10.

4. Flattening of an image destroys the spatial relationship between input features. The model learns these relationships from the flattened one-dimensional vector. We will see later how CNNs (convolutional neural networks) can be used to overcome this shortcoming.

5.  Loss and metric (sparse categorical accuracy) are
    plotted in Figures 2-46 and 2-47.

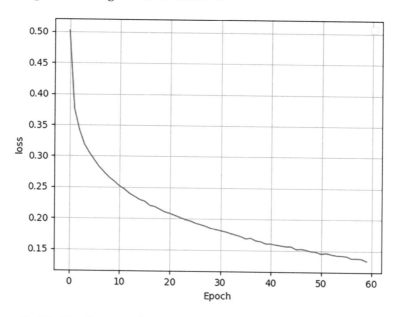

***Figure 2-46.*** *Evolution of Sparse Categorical Cross Entropy Loss with Epochs*

6.  The model achieves about 95% accuracy as can be
    seen from the confusion matrix for training data in
    Figure 2-48. The confusion matrix for testing data is
    shown in Figure 2-49. The plots show that the model
    classifies the vast majority of images correctly.

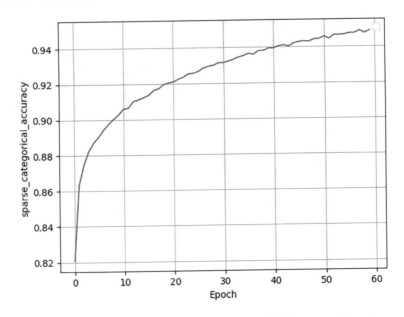

**Figure 2-47.** *Evolution of Sparse Categorical Accuracy Metric with Epochs*

The complete code for fashion MNIST image classification is shown in Listing 2-31.

**Listing 2-31.** Classifying Images from the Fashion MNIST Dataset

```
1    import numpy as np
2    import tensorflow as tf
3    import matplotlib.pyplot as plt
4    import logging
5    import os
6    import seaborn as sns
7
8    logging.basicConfig(level=logging.DEBUG)
9
10
11   class History(object):
```

```
12      def __init__(self):
13          self.history = {}
14          self.epoch = None
15
16
17  class FashionMNistClassify(object):
18      LOGGER = logging.getLogger("FashionMNistClassify")
19
20      def __init__(self, datadir: str, batchsize: int = 10,
            epochs: int = 20, useGradTape: bool = True) -> None:
21          (trainx, trainy), (testx, testy) = tf.keras.
                datasets.fashion_mnist.load_data()
22          self.classes = ["Top", "Trouser", "Pullover",
                "Dress", "Coat", "Sandal", "Shirt", "Sneaker",
23                              "Bag", "Boot"]
24          self.nClass = len(self.classes)
25          trainx = trainx/255.0
26          testx = testx/255.0
27          self.trainingData = (trainx, trainy)
28          self.testingData = (testx, testy)
29          self.inputDir = datadir
30          self.batchSize = batchsize
31          self.nEpoch = epochs
32          self.useGradientTape = useGradTape
33          self.nnet = self.model()
34
35      def checkpointModel(self, nnet):
36          checkpointFile = os.path.join(self.inputDir,
                "checkpoint_fmnist_wt")
37          if not os.path.exists(checkpointFile):
```

```
38            nnet.predict(np.ones((20, 28, 28), dtype=np.
              float32))
39            tf.keras.models.save_model(nnet,
              checkpointFile, overwrite=False)
40        else:
41            nnet = tf.keras.models.load_
              model(checkpointFile)
42        return nnet
43
44    def model(self):
45        nnet = tf.keras.models.Sequential()
46        nnet.add(tf.keras.layers.Flatten(input_shape=
          (28, 28)))
47        nnet.add(tf.keras.layers.Dense(80,
          activation="relu"))
48        nnet.add(tf.keras.layers.Dense(20, activation=
          "relu"))
49        nnet.add(tf.keras.layers.Dense(10))
50        self.loss = tf.keras.losses.SparseCategoricalCros
          sentropy(from_logits=True)
51        self.optimizer = tf.keras.optimizers.Adam
          (learning_rate=0.005)
52        self.metric = tf.keras.metrics.SparseCategorical
          Accuracy()
53        nnet.compile(optimizer=self.optimizer,
54                     loss=self.loss,
55                     metrics=[self.metric])
56        nnet = self.checkpointModel(nnet)
57        return nnet
58
```

```
59      def plotConfusionMatrix(self, labels: np.ndarray,
        predictions: np.ndarray) -> None:
60          predictedLabels = np.argmax(predictions, axis=1)
61          fig, ax = plt.subplots()
62          cm = np.zeros((self.nClass, self.nClass),
            dtype=np.int32)
63          for i in range(labels.shape[0]):
64              cm[labels[i], predictedLabels[i]] += 1
65          sns.heatmap(cm, annot=True, fmt="d",
            linewidths=0.25, ax=ax)
66          ax.set_xticks(range(1+self.nClass))
67          ax.set_yticks(range(1+self.nClass))
68          ax.set_xticklabels(["0"] + self.classes,
            rotation=20)
69          ax.set_yticklabels(["0"] + self.classes,
            rotation=20)
70          ax.set_ylabel('Actual')
71          ax.set_xlabel('Predicted')
72          plt.show()
73
74      def plotConvergenceHistory(self, history,
        metricName):
75          plt.plot(history.epoch, history.
            history[metricName])
76          plt.xlabel("Epoch")
77          plt.ylabel(metricName)
78          plt.grid(True)
79          plt.legend()
80          plt.show()
81
82      def testModel(self):
```

```
83              for X, y in [self.trainingData, self.
                testingData]:
84                  predictClass = self.nnet.predict(X)
85                  self.plotConfusionMatrix(y, predictClass)
86
87          def gradTapeTraining(self):
88              trainDataset = tf.data.Dataset.from_tensor_
                slices(self.trainingData)
89              trainDataset = trainDataset.batch(self.batchSize)
90              totalLoss = np.zeros(self.nEpoch, dtype=np.
                float32)
91              count = 0
92              for X, y in trainDataset:
93                  for epoch in range(self.nEpoch):
94                      with tf.GradientTape() as tape:
95                          predictedY = self.nnet(X)
96                          loss = self.loss(y, predictedY)
97
98                      grads = tape.gradient(loss, self.nnet.
                        trainable_weights)
99                      self.LOGGER.info("Epoch %d, loss %f",
                        epoch, loss)
100                     totalLoss[epoch] += loss
101                     self.optimizer.apply_gradients(zip(grads,
                        self.nnet.trainable_weights))
102                 count += 1
103             totalLoss = totalLoss / count
104             history = History()
105             history.history["loss"] = totalLoss
106             history.history[self.metric._name] =
                np.zeros(self.nEpoch)
```

```
107            history.epoch = np.arange(self.nEpoch)
108            return history
109
110        def trainModel(self):
111            if self.useGradientTape:
112                history = self.gradTapeTraining()
113            else:
114                history = self.nnet.fit(self.trainingData[0],
                       self.trainingData[1],
115                                          batch_size=self.
                                          batchSize,
                                          epochs=self.nEpoch)
116            self.plotConvergenceHistory(history, self.
               metric._name)
117            self.plotConvergenceHistory(history, "loss")
118            return history
119
120
121    if __name__ == "__main__":
122        dname = r"C:\prog\cygwin\home\samit_000\RLPy\
           data\book"
123        fmnist = FashionMNistClassify(dname, batchsize=10000,
           epochs=60)
124        fmnist.trainModel()
125        fmnist.testModel()
```

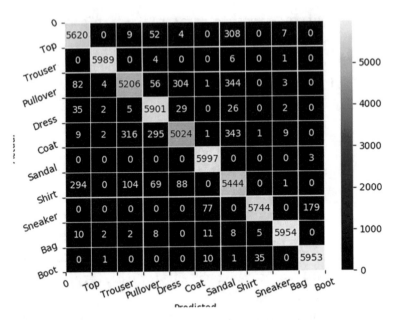

***Figure 2-48.*** *Confusion Matrix of Model Predictions on the Training Dataset*

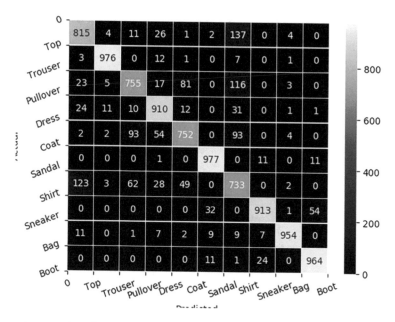

***Figure 2-49.*** *Confusion Matrix of Model Predictions on the Testing Dataset*

# CHAPTER 3

# Convolutional Neural Networks

Convolutional neural networks (CNNs) are a category of neural networks that can be used to identify spatial patterns in a robust manner. They achieve this robustness by a parsimonious use of parameters and by systematic identification of simple patterns that are aggregated into complex specifications using subsequent layers. They are able to recognize features in a translation-invariant fashion. Their use in modern digital imaging technology is ubiquitous, with most cameras programmed to pick out faces and objects automatically using CNNs.

The development of CNNs was inspired by the human visual cortex. The human eye can recognize an object at different positions in its field of view. In 1959, David Hubel and Torsten Wiesel proposed a theory to account for the spatial invariance of the human eye's object detection ability by surmising that the human eye has simple and complex cells, with simple cells tracking the presence of an object at a particular location and complex cells aggregating the output of simple cells. In 1980, Dr. Kunihiko Fukushima implemented a neural network model (neurocognitron model) to simulate the functioning of the human eye as described by Hubel and Weisel. The first notable application of convolutional neural networks to the task of pattern recognition came in 1998, when LeCun, Bottou, Bengio, and Haffner used a CNN to identify handwritten digits from the MNIST database. The convolutional neural network designed by LeCun

S. Ahlawat, *Reinforcement Learning for Finance*,
https://doi.org/10.1007/978-1-4842-8835-1_3

et al. was called LeNet. The next major breakthrough in the field came in 2012 when a group of researchers from the University of Toronto, led by Alex Krizhevsky, created an image recognition algorithm that achieved 85% accuracy. The adoption of CNNs gained further traction in the field of image recognition with access to ever-growing computational resources. In the past few years, CNNs have been used for face recognition, fingerprint recognition, medical image analysis, and motion detection with increasing accuracy.

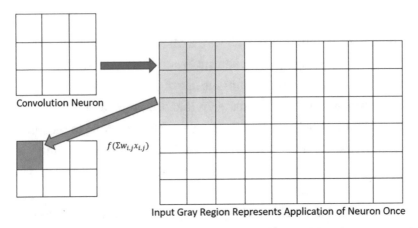

Input Gray Region Represents Application of Neuron Once

**Two-Dimensional 3X3 Convolution Neuron with Stride 3X3 Applied to 6X9 Input to get 2X3 Output**

***Figure 3-1.*** *Application of a CNN Neuron to Inputs*

## 3.1  A Simple CNN

A convolutional neural network consists of one or more convolution layers. A convolution layer is comprised of a set of neurons. What distinguishes a neuron in a convolution layer from a neuron in a dense layer is the shape of its input and the method of applying inputs to activate the neuron. A convolution neuron is applied repeatedly to different sections of the input in order to obtain the output. For example, Figure 3-1 shows a two-dimensional convolution neuron with size 3 by 3, applied to a two-dimensional input. The output of this neuron is $f\left(\sum_{i,j} W_{i,j} X_{i,j} + b\right)$, where $f$ is the activation function, $W_{i,j}$ represents neuron weights, $X_{i,j}$ is the input, and $b$ is the bias. Let us look at an example using TensorFlow.

***Listing 3-1.*** Example of a 2D Convolutional Neural Network

```
1    from tensorflow.keras import layers, models
2
3    model = models.Sequential()
4    model.add(layers.Conv2D(10, (3, 3), activation="relu",
     input_shape=(20, 20, 1)))
5    model.add(layers.MaxPooling2D((2, 2)))
6    model.add(layers.Conv2D(20, (3, 3), activation="relu"))
7    model.add(layers.MaxPooling2D((2, 2)))
8    model.add(layers.Conv2D(20, (3, 3), activation="relu"))
9    model.add(layers.Flatten())
10   model.add(layers.Dense(16, activation="relu"))
11   model.add(layers.Dense(1, activation="sigmoid"))
12   print(model.summary())
```

A sequential neural network is built by adding layers. In the example shown in Listing 3-1, the first layer is a two-dimensional convolution layer that has ten neurons, each having a two-dimensional kernel of length and width 3. Each neuron uses a ReLU activation function (given in equation 3.1). The rectified linear unit or ReLU activation function does not get saturated for large or small activation values, unlike sigmoid or hyperbolic functions that asymptotically saturate to a value of 1 for large activation and 0 or –1 for small activation. Being the first layer of the network, the convolution layer also needs input shape specification. For a two-dimensional convolution layer, the shape must have three dimensions: length, width, and height. In this example, the convolution neuron accepts an input of size (3, 3, 1). The last 1 comes from the last dimension of the input. For example, if we were using RGB images as input, the last dimension of input would be 3. Consequently, for an input shape of (20, 20, 3), the convolution neuron would accept an input of size (3, 3, 3). In general, for a two-dimensional convolution layer shown in the following with $N$ neurons or filters, each with a filter size of $L$ by $W$, applied on an input shape of $I_L$ by $I_W$ by $H$, each neuron will accept an input of dimension $L$ by $W$ by $H$. The default activation function for a layer is identity, that is, an output equal to the input.

$$y = max(0, x) \tag{3.1}$$

*Listing 3-2.* 2D Convolution Layer

```
1    from tensorflow.keras import layers
2    layers.Conv2D(N, (L, W), strides=(strideX, strideY),
3                                activation="relu",
4                                input_shape=(iL, iW, H) )
```

The default stride length of a two-dimensional convolution neuron is 1 by 1. In the example shown in Listing 3-1, applying stride of 1 to input of length 20, with filter length 3, will give $20 - 3 + 1 = 18$ outputs. The same convolution neuron is applied to different portions of the input, yielding an output array of shape 18 by 18. Because there are ten neurons in the first layer, output shape is 18 by 18 by 10. For the generic example shown in Listing 3-2, output shape is given by equation 3.2. The number of parameters in this layer is given by equation 3.3, where $N$ is the number of neurons in the convolution layer, $L$ by $W$ is the filter size, and $H$ is the height. A neuron has a weight for each of the $L$ by $W$ by $H$ inputs in addition to bias. For the first layer in Listing 3-1, it has $(3 \times 3 \times 1 + 1) \times 10 = 100$ parameters.

$$S = \frac{I_L - L}{stride\ X} + 1 \qquad (3.2)$$

$$P = (L \times W \times H + 1) \times N \qquad (3.3)$$

TensorFlow follows the convention of using the first dimension as the number of batches. For example, if the data being fed to a CNN has 50 specimens, the first dimension of the input would be 50. This allows the input to be sent in one shot and facilitates greater execution efficiency due to batch processing. The output shapes of layers shown by the summary command show the first output dimension as None, signifying it is the number of specimens from the input, as shown in Listing 3-3.

***Listing 3-3.*** Model Summary

```
1   Model: "sequential_2"
2   _____
3   Layer (type)                    Output Shape              Param #
4   _____
5   conv2d_3 (Conv2D)               (None, 18, 18, 10)        100
6   _____
7   max_pooling2d (MaxPooling2D)    (None, 9, 9, 10)          0
8   _____
9   conv2d_4 (Conv2D)               (None, 7, 7, 20)          1820
10  _____
11  max_pooling2d_1 (MaxPooling2    (None, 3, 3, 20)          0
12  _____
13  conv2d_5 (Conv2D)               (None, 1, 1, 20)          3620
14  _____
15  flatten (Flatten)               (None, 20)                0
16  _____
```

| 17 | dense (Dense) | (None, 16) | 336 |
|----|----|----|----|
| 18 | | | |
| 19 | dense_1 (Dense) | (None, 1) | 17 |
| 20 | | | |
| 21 | Total params: 5,893 | | |
| 22 | Trainable params: 5,893 | | |
| 23 | Non-trainable params: 0 | | |
| 24 | | | |

A three-dimensional convolution layer is created using **layers. Conv3D**. Similar to its two-dimensional counterpart, this layer takes the number of neurons and a three-dimensional filter size as shown in Listing 3-4. Input shape must be a four-element tuple. Each neuron accepts input of shape $L$ by $W$ by $H$ by $N_{channels}$. Similarly, a one-dimensional convolution layer takes a one-dimensional filter size, one-dimensional stride, and two-dimensional input, as shown in Listing 3-5.

***Listing 3-4.*** 3D Convolution Layer

```
1   from tensorflow.keras import layers
2   layers.Conv3D(N, (L, W, H),
3                 strides=(strideX, strideY, strideZ),
4                 activation="relu",
5                 input_shape=(I_L, I_W, I_H, nChannels) )
```

***Listing 3-5.*** 1D Convolution Layer

```
1   from tensorflow.keras import layers
2   layers.Conv1D(N, (L), strides=(strideX),
3                 activation="relu",
4                 input_shape=(I_L, nChannels) )
```

The next layer in Listing 3-1 is a pooling layer and aggregates information from neighboring cells. A max pooling layer takes the maximum value of its constituent cells and has no parameters. It is applied in a nonoverlapping fashion over the input. The input for this layer is the previous layer's output, having a shape of $(None, 18, 18, 10)$. Since this is a two-dimensional pooling layer with shape 2 by 2, it produces an output of shape $(None, 9, 9, 10)$, with the length and width of the output reduced by a factor corresponding to the pooling layer's length and width.

The third layer is another two-dimensional convolution layer. This layer recognizes features from the aggregated output of the max pooling layer. By progressively recognizing complex patterns from simpler building block patterns, a convolutional neural network can identify complex features. This layer has 20 neurons, each with a filter size of 3 by 3. Using equation 3.2, this gives an output of size $(None, 7, 7, 20)$. The number of parameters is $(3 \times 3 \times 10 + 1) \times 20 = 1820$, using equation 3.3.

The next two layers follow a similar pattern: a max pooling layer followed by a two-dimensional convolution layer. The max pooling layer has no free parameter, and the shape of its output is $(None, 3, 3, 30)$. The max pooling layer operates on an input of shape $(None, 7, 7, 20)$ from the last layer. The filter size of the max pooling layer is $(2, 2)$, but the input shape $(7, 7)$ is odd. This causes the last row and column of the input to be discarded. This situation is not ideal, as it leads to information loss. In order to fix the issue, a modified CNN is shown in Listing 3-6. An alternative fix for this problem would be to use padding. Padding appends additional rows or columns to the input with value $-\infty$ for a max pooling layer and 0 for an average pooling layer. The **Padding** argument accepts one of two values: **valid** or **same**. **valid** padding is the default selection for padding and ignores sections of input not covered by a complete pooling layer's span. '*same*' applies padding to the input, so that no section of input is ignored by the pooling layer. For a max pooling layer, it applies $-\infty$ padded values so that the padded output has no impact on actual output. Similarly, an average pooling layer applies a padding of 0, leaving the output unaltered by padded values. The pooling layer also takes an optional argument specifying the stride. It defaults to the filter size.

The next convolution layer has $(3 \times 3 \times 20 + 1) \times 20 = 3620$ free parameters and output of shape (*None*, 1, 1, 20). Finally, output from this layer is fed into a flatten layer that changes the input to a one-dimensional input of shape (*None*, 20). Following the convention, the first dimension is reserved for batch size, and the second dimension is $1 \times 1 \times 20 = 20$, from the output shape of the last layer. The output from the flatten layer is fed to a dense layer with 16 neurons. This layer has $(20 + 1) \times 16 = 336$ free parameters. In general, the number of parameters of a dense layer is shown in equation 3.4, where $N_{input}$ is the number of inputs to the layer and $N$ is the number of neurons in the layer. 1 accounts for bias weight.

The final layer in Listing 3-1 is a one-neuron dense layer with a sigmoid activation function. This layer produces a scalar output between 0 and 1 and can be interpreted as the probability of belonging to a class. The number of free parameters for this layer is $(16 + 1) \times 1 = 17$, with the output shape being (*None*, 1).

$$P = \left( N_{input} + 1 \right) \times N \tag{3.4}$$

**Listing 3-6.** Example of a 2D Convolutional Neural Network with No Data Loss in the Max Pooling Layer

```
1    import tensorflow as tf
2    from tensorflow.keras import layers, models
3
4    model = models.Sequential()
5    model.add(layers.Conv2D(10, (5, 5), activation="relu",
     input_shape=(20, 20, 1)))
```

```
 6    model.add(layers.MaxPooling2D((2, 2)))
 7    model.add(layers.Conv2D(20, (3, 3), activation="relu"))
 8    model.add(layers.MaxPooling2D((2, 2)))
 9    model.add(layers.Conv2D(20, (3, 3), activation="relu"))
10    model.add(layers.Flatten())
11    model.add(layers.Dense(16, activation="relu"))
12    model.add(layers.Dense(1), activation="sigmoid")
13    model.summary()
14
15    Model: "sequential"
16
```

| Layer (type) | Output Shape | Param # |
|---|---|---|
| conv2d (Conv2D) | (None, 16, 16, 10) | 260 |
| max_pooling2d (MaxPooling2D) | (None, 8, 8, 10) | 0 |
| conv2d_1 (Conv2D) | (None, 6, 6, 20) | 1820 |
| max_pooling2d_1 (MaxPooling2 | (None, 3, 3, 20) | 0 |
| conv2d_2 (Conv2D) | (None, 1, 1, 20) | 3620 |
| flatten (Flatten) | (None, 20) | 0 |
| dense (Dense) | (None, 16) | 336 |
| dense_1 (Dense) | (None, 1) | 17 |

```
35    Total params: 6,053
```

```
36    Trainable params: 6,053
37    Non-trainable params: 0
38 _____
```

# 3.2  Neural Network Layers Used in CNNs

In order to identify spatial patterns, convolutional neural networks frequently use the following layers. For all convolution layers, the depth of input matches the depth of the filter:

1. **One-dimensional convolution layer**: In TensorFlow, this is defined in class **tf.keras.layers. Conv1D**. It takes the number of filters and kernel size as arguments. Stride along one dimension can be specified. Kernel size is an integer representing the length of the filter or a tuple containing filter length and depth.

2. **Two-dimensional convolution layer**: Defined in TensorFlow class **tf.keras.layers.Conv2D**, this layer takes the number of filters and kernel size as arguments. Stride along two dimensions can be specified as a tuple. Kernel size is a tuple specifying the length and the width of the filter. The height of the filter matches the height (depth) of the input. This is the fourth dimension of input.

3. **Three-dimensional convolution layer**: Defined in TensorFlow class **tf.keras.layers.Conv3D**, this layer takes the number of filters and kernel size as arguments. Stride along three dimensions can be specified as a tuple. Kernel size is a tuple specifying the length, width, and height of the filter.

4. **One-dimensional convolutional transpose layer**: This layer applies inverse convolution (deconvolution) transformation, taking the output of a one-dimensional convolution layer as input and producing an output with shape corresponding to the original input to the convolution layer. In TensorFlow, this is defined in class **tf.keras.layers. Conv1DTranspose**.

5. **Two-dimensional convolutional transpose layer**: This layer is defined in TensorFlow class **tf.keras.layers.Conv2DTranspose** and is the two-dimensional equivalent of the Conv1DTranspose layer. A code example illustrating its use is shown in Listing 3-7. In this example, input height is 3 and is also the filter height.

***Listing 3-7.*** Example of a 2D Convolutional Transpose Layer

```
1  import tensorflow as tf
2  convLayer = tf.keras.layers.Conv2D(10, (4, 4),
   strides=(2, 2), kernel_initializer="ones",
3                                  bias_initializer=
                                   "ones", input_shape=(8,8,3))
4  deconvLayer = tf.keras.layers.Conv2DTranspose(3, (4, 4),
   strides=(2,2),
5                     kernel_initializer=tf.keras.
                      initializers.Constant(1.0/(49*4)),
6                         bias_initializer="ones")
7  input = tf.constant(tf.ones((1, 8, 8, 3), dtype=tf.
   float32))
```

```
8    out1 = convLayer(input)
9    out2 = deconvLayer(out1)
10   assert out2.shape == input.shape
```

6. **Three-dimensional convolutional transpose layer**: Defined in TensorFlow class **tf.keras.layers. Conv3DTranspose**, this layer applies three-dimensional deconvolution.

# 3.3 Output Shapes and Trainable Parameters of CNNs

Having familiarized ourselves with CNN terminology, we are now in a position to formulate mathematical expressions for output shape and number of trainable parameters of a CNN layer.

Let us consider a general three-dimensional CNN and denote the number of filters (neurons) by $N$, filter shape by $(L, W, D)$, input shape by $(B, I_L, I_W, I_D, H)$, and stride as $(S_L, S_W, S_D)$. $D$ represents the depth, $L$ length, and $W$ width of the filter. $B$ is the number of batches in input, $I_L$ is input length, $I_W$ is input width, and $I_D$ is input depth. $H$ is input height and is equal to filter height. Filter height is not provided as an input in filter shape because it is automatically set to match input height. Let us assume a padding of shape $(P_L, P_W, P_D)$ on both sides of length, width, and depth of input. Output shape of the CNN layer is shown in equation 3.5. The expression can be understood as moving a filter of length $L$ along $I_L$ padded with $P_L$ on both sides with stride $S_L$ will result in the top-left corner of the filter traveling from 0 to $I_L - L + 2P_L + 1$ with stride of $S_L$ giving the output length $(B, O_L, O_W, O_D, N)$ as shown in equation 3.5:

$$B = \text{number of batches}$$

$$O_L = \frac{I_L - L + 2P_L}{S_L} + 1$$

$$O_W = \frac{I_W - W + 2P_W}{S_W} + 1 \tag{3.5}$$

$$O_D = \frac{I_D - D + 2P_D}{S_D} + 1$$

$$N = \text{number of filters}$$

The number of trainable parameters of a CNN layer is determined by its filter size, inclusive of filter height, and bias. Since filter height is the same as input height, the number of trainable parameters is given by equation 3.6. 1 accounts for bias weight:

$$N_{param} = (L \times W \times D \times H + 1)N \tag{3.6}$$

Similarly, for a two-dimensional CNN, output shape and number of trainable parameters are given by equations 3.7 and 3.8, respectively:

$$B = \text{number of batches}$$

$$O_L = \frac{I_L - L + 2P_L}{S_L} + 1$$

$$O_W = \frac{I_W - W + 2P_W}{S_W} + 1 \tag{3.7}$$

$$N = \text{number of filters}$$

$$N_{param} = (L \times W \times H + 1)N \tag{3.8}$$

One can write analogous expressions for output shape and number of trainable parameters for a one-dimensional CNN layer.

# 3.4 Classifying Fashion MNIST Images

In Chapter 2, a simple neural network model using dense layers was built to classify images in the fashion MNIST dataset with ≈95% accuracy on training data. In this section, let us build a CNN-based network to improve the performance of the image classifier. The complete code is shown in Listing 3-8. The directory specified on line 91 should be changed to a writable directory and is used for creating a model checkpoint.

1. Two-dimensional convolution layers are used to detect spatial features in an image. As seen from code in Listing 3-8, a first CNN layer must specify input shape as a three-element tuple: image length, image width, and image depth. In this example, images use gray scale, and there is no RGB channel. So depth is 1.

2. A max pooling layer is used to aggregate patterns.

3. Output from a convolution layer is flattened and sent to a dense layer.

4. A final dense layer has ten units, each predicting the unnormalized probability of an image belonging to that class. To recall, this problem has ten classes.

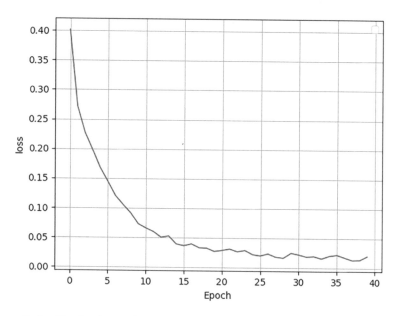

***Figure 3-2.*** *Evolution of Sparse Categorical Cross Entropy Loss with Epochs Using CNN*

5.  Plots for evolution of loss and accuracy shown in Figures 3-2 and 3-3 demonstrate that training has converged.

6.  The CNN-based model has accuracy of 99.5% in the training dataset and 90.6% in the testing dataset. Confusion matrices plotting actual labels (Y axis) vs. predicted labels (X axis) for training and testing datasets can be seen in Figures 3-4 and 3-5.

***Listing 3-8.*** Classifying Images from the Fashion MNIST Dataset Using CNN

```
1   import numpy as np
2   import tensorflow as tf
3   import matplotlib.pyplot as plt
```

```
4    import logging
5    import os
6    import seaborn as sns
7
8    logging.basicConfig(level=logging.DEBUG)
9
10
11   class FashionMNistCNNClassify(object):
12       LOGGER = logging.getLogger("FashionMNistCNNClassify")
13
14       def __init__(self, datadir: str, batchsize: int = 10,
         epochs: int = 20) -> None:
15           (trainx, trainy), (testx, testy) = tf.keras.
             datasets.fashion_mnist.load_data()
16           self.classes = ["Top", "Trouser", "Pullover",
             "Dress", "Coat", "Sandal", "Shirt", "Sneaker",
17                               "Bag", "Boot"]
18           self.nClass = len(self.classes)
19           trainx = trainx/255.0
20           testx = testx/255.0
21           self.trainingData = (trainx, trainy)
22           self.testingData = (testx, testy)
23           self.inputDir = datadir
24           self.batchSize = batchsize
25           self.nEpoch = epochs
26           self.nnet = self.model()
27
28       def checkpointModel(self, nnet):
29           checkpointFile = os.path.join(self.inputDir,
             "checkpoint_fmnist_cnn_wt")
30           if not os.path.exists(checkpointFile):
```

```
31          nnet.predict(np.ones((20, 28, 28, 1),
            dtype=np.float32))
32          tf.keras.models.save_model(nnet,
            checkpointFile, overwrite=False)
33      else:
34          nnet = tf.keras.models.load_
            model(checkpointFile)
35      return nnet
36
37  def model(self):
38      nnet = tf.keras.models.Sequential()
39      nnet.add(tf.keras.layers.Conv2D(filters=100,
        kernel_size=(2, 2), padding="same", input_
        shape=(28, 28, 1)))
40      nnet.add(tf.keras.layers.MaxPooling2D(pool_
        size=(2, 2)))
41      nnet.add(tf.keras.layers.Conv2D(filters=60,
        kernel_size=(2, 2), padding="same",
        activation="relu"))
42      nnet.add(tf.keras.layers.Flatten())
43      nnet.add(tf.keras.layers.Dense(50,
        activation="relu"))
44      nnet.add(tf.keras.layers.Dense(10))
45      self.loss = tf.keras.losses.SparseCategoricalCross
        entropy(from_logits=True)
46      self.optimizer = tf.keras.optimizers.
        Adam(learning_rate=0.002)
47      self.metric = tf.keras.metrics.
        SparseCategoricalAccuracy()
48      nnet.compile(optimizer=self.optimizer,
49                   loss=self.loss,
```

```
50                        metrics=[self.metric])
51          nnet = self.checkpointModel(nnet)
52          return nnet
53
54      def plotConfusionMatrix(self, labels: np.ndarray,
        predictions: np.ndarray) -> None:
55          predictedLabels = np.argmax(predictions, axis=1)
56          fig, ax = plt.subplots()
57          cm = np.zeros((self.nClass, self.nClass),
            dtype=np.int32)
58          for i in range(labels.shape[0]):
59              cm[labels[i], predictedLabels[i]] += 1
60          sns.heatmap(cm, annot=True, fmt="d",
            linewidths=0.25, ax=ax)
61          ax.set_xticks(range(1+self.nClass))
62          ax.set_yticks(range(1+self.nClass))
63          ax.set_xticklabels(["0"] + self.classes,
            rotation=20)
64          ax.set_yticklabels(["0"] + self.classes,
            rotation=20)
65          ax.set_ylabel('Actual')
66          ax.set_xlabel('Predicted')
67          plt.show()
68
69      def plotConvergenceHistory(self, history, metricName):
70          plt.plot(history.epoch, history.
            history[metricName])
71          plt.xlabel("Epoch")
72          plt.ylabel(metricName)
73          plt.grid(True)
74          plt.legend()
```

```
75          plt.show()
76
77      def testModel(self):
78          for X, y in [self.trainingData, self.testingData]:
79              predictClass = self.nnet.predict(X[...,
                np.newaxis])
80              self.plotConfusionMatrix(y, predictClass)
81
82      def trainModel(self):
83          history = self.nnet.fit(self.trainingData[0][...,
            np.newaxis], self.trainingData[1],
84                          batch_size=self.batchSize,
                            epochs=self.nEpoch)
85          self.plotConvergenceHistory(history, self.
            metric._name)
86          self.plotConvergenceHistory(history, "loss")
87          return history
88
89
90  if __name__ == "__main__":
91      dname = r"C:\prog\cygwin\home\samit_000\RLPy\
        data\book"
92      fmnist = FashionMNistCNNClassify(dname, batchsize=100,
        epochs=40)
93      fmnist.trainModel()
94      fmnist.testModel()
```

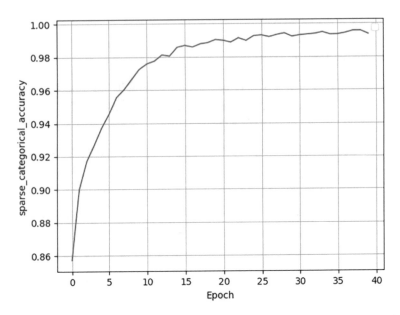

**Figure 3-3.** *Evolution of Sparse Categorical Accuracy Metric with Epochs Using CNN*

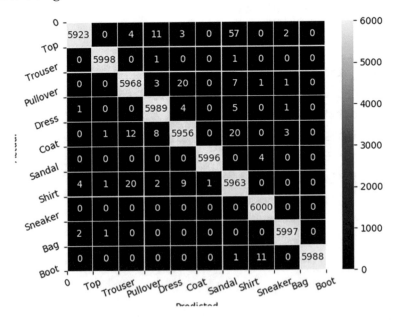

**Figure 3-4.** *Confusion Matrix of CNN Model Predictions (X Axis) Against Actual Labels (Y Axis) on the Training Dataset*

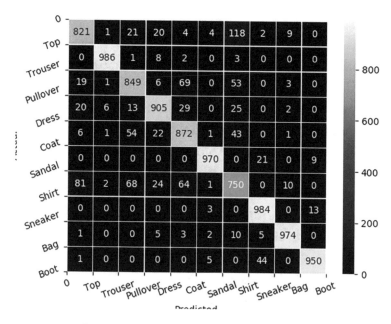

***Figure 3-5.***  *Confusion Matrix of CNN Model Predictions (X) Against Actual Labels (Y) on the Testing Dataset*

# 3.5  Identifying Technical Patterns in Security Prices

This section applies convolutional neural networks to the task of pattern recognition in security prices. Technical patterns are widely used in securities markets for arbitrage and risk management. Trading firms track the moving averages of security prices to figure out when to initiate a long position in a stock. Cup-and-handle, moving average crossover rule, head-and-shoulders, rising wedge, and falling wedge are a few examples of widely used technical patterns to predict the future course of price movements. Pattern recognition shares similarities and differences with image recognition. Like images, patterns are graphical representations whose essential features need to be learned by a network. Unlike image recognition, their presence may be obscured by daily price fluctuations.

While one level of smoothing may reveal a pattern to the eye in one period, another level of smoothing may be needed to detect its existence in another period. Furthermore, technical analysts may differ in their opinion on occurrence or non-occurrence of a pattern depending upon feature sizes. For example, some technical analysts insist that the length of the handle in the cup-and-handle pattern should be at least *one – third* the length of the entire pattern, while other analysts disagree.

In financial applications, data is finite, and models must account for available training data when deciding the number of model parameters. Judicious selection of the number of model parameters is critical for another reason – to avoid overfitting. We want the model to learn underlying features of a pattern, without learning the noise. With these considerations in mind, let us use CNNs to detect the occurrence of the cup-and-handle price pattern in the security prices of five stocks from 2000 to 2020. The CNN used for pattern identification is shown in Listing 3-9.

***Listing 3-9.*** 2D Convolutional Neural Network for Identifying the Cup-and-Handle Pattern

```
1   from tensorflow.keras import layers, models
2
3   model = models.Sequential()
4   model.add(layers.Conv2D(10, (3, 3), activation="relu",
    input_shape=(20, 20, 1)))
5   model.add(layers.AveragePooling2D(pool_size=(2, 2)))
6   model.add(layers.Conv2D(5, (4, 4), activation="relu"))
7   model.add(layers.Flatten())
8   model.add(layers.Dense(2, activation="relu"))
9   model.add(layers.Dense(1, activation="sigmoid"))
10  print(model.summary())
11
12  Model: "sequential_1"
```

| Layer (type) | Output Shape | Param # |
|---|---|---|
| conv2d_2 (Conv2D) | (None, 18, 18, 10) | 100 |
| average_pooling2d_1 (Average | (None, 9, 9, 10) | 0 |
| conv2d_3 (Conv2D) | (None, 6, 6, 5) | 805 |
| flatten_1 (Flatten) | (None, 180) | 0 |
| dense_2 (Dense) | (None, 2) | 362 |
| dense_3 (Dense) | (None, 1) | 3 |

```
Total params: 1,270
Trainable params: 1,270
Non-trainable params: 0
```

A known cup-and-handle pattern in the price of a security was identified. This pattern manifested itself in the price of BIDU from February 1, 2007, to May 3, 2007. The price plot is illustrated in Figure 3-6. In order to generate a sufficient number of testing samples containing both occurrences and non-occurrences of this pattern for training the CNN with 1,270 parameters, more testing data is required. Furthermore, manual identification and confirmation of the occurrence or non-occurrence of the pattern would be too cumbersome. In order to overcome this problem, small random noise was added to the price of this security in the period of interest when it showed a confirmed occurrence of the cup-and-handle pattern. The random disturbance was produced using a Gaussian distribution with 0 mean and standard deviation 0.2. Since the cup-and-handle pattern spans a price range of around $3.5, the noise has a small enough standard deviation, and it's addition to the price is unlikely to negate the occurrence of the cup-and-handle pattern. With a certainty of 99.73%, the disturbance will be between three standard deviations, or between [−0.6,0.6]. And in this range, none of the essential features of the cup-and-handle pattern will be obscured.

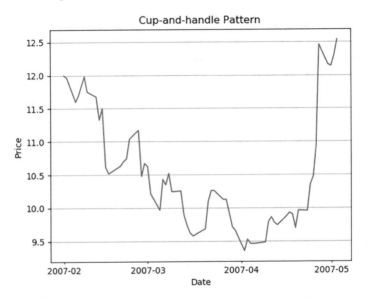

***Figure 3-6.*** *Cup-and-Handle Pattern*

Finding data with no cup-and-handle pattern is easier. The price of a security was considered in a span of 3 months and tested for essential features of the cup-and-handle pattern. If the pattern is not detected, which is a fairly common occurrence, it is added to the training data as a negative sample.

Armed with sufficient data comprising both positive and negative samples for training the CNN, the network was trained and validated. The trained network was used to identify the occurrences of the cup-and-handle pattern in prices of securities. A few positive occurrences are displayed in Figures 3-7 and 3-8. As can be seen, the network can identify the pattern.

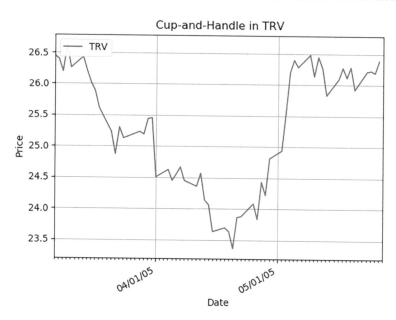

***Figure 3-7.*** *Cup-and-Handle Pattern in the TRV Stock Price Chart*

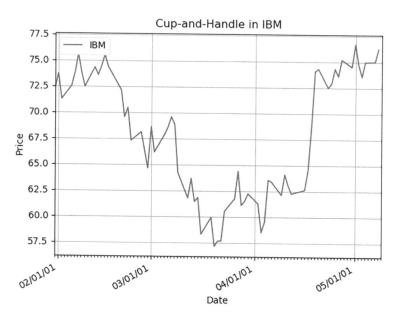

***Figure 3-8.*** *Cup-and-Handle Pattern in the IBM Stock Price Chart*

The full code for cup-and-handle pattern recognition in stock prices of 30 Dow Jones components is presented in Listing 3-10.

***Listing 3-10.*** Cup-and-Handle Pattern Recognition in Prices of 30 Dow Jones Components

```
1    import numpy as np
2    import pandas as pd
3    import tensorflow as tf
4    from tensorflow.keras import layers, models
5    import os
6
7    import matplotlib.pyplot as plt
8    from matplotlib.dates import (YEARLY, DateFormatter,
9                                  YearLocator, MonthLocator,
                                    DayLocator)
10
11
12   class DatePlotter(object):
13       def __init__(self):
14           self.majorLocator = MonthLocator()
             #YearLocator()  # every year
15           self.minorLocator = DayLocator()  # every month
16           self.formatter = DateFormatter('%m/%d/%y')
             #DateFormatter('%Y')
17
18       def plot(self, df, datecol, valcols, xlabel='date',
         ylabel=None, labels=None, round='Y'):
19           if not labels:
20               labels = valcols
21
22           fig, ax = plt.subplots()
```

```
23        for val,lb in zip(valcols, labels):
24            ax.plot(datecol, val, data=df, label=lb)
25        plt.xlabel(xlabel)
26        if ylabel:
27            plt.ylabel(ylabel)
28
29        # format the ticks
30        ax.xaxis.set_major_locator(self.majorLocator)
31        ax.xaxis.set_major_formatter(self.formatter)
32        ax.xaxis.set_minor_locator(self.minorLocator)
33
34        # round to nearest years.
35        datemin = np.datetime64(df[datecol].
          values[0], round)
36        nr = df.shape[0]-1
37        datemax = np.datetime64(df[datecol].values[nr],
          round) + np.timedelta64(1, round)
38        ax.set_xlim(datemin, datemax)
39
40        # format the coords message box
41        ax.format_xdata = DateFormatter('%Y-%m-%d')
42        ax.format_ydata = lambda x: '$%1.2f' % x  #
          format the price.
43        handles, labels = ax.get_legend_handles_labels()
44        ax.legend(handles, labels, loc='upper left')
45        ax.grid(True)
46
47        # rotates and right aligns the x labels, and
          moves the bottom of the
48        # axes up to make room for them
49        fig.autofmt_xdate()
```

```
50          return plt
51
52
53   def plotData(price_data_dir, output_dir):
54       df = pd.read_csv(os.path.join(output_dir, "ch_
         out.csv"))
55       df.loc[:, "Begin"] = pd.to_datetime(df.loc[:,
         "Begin"])
56       df.loc[:, "End"] = pd.to_datetime(df.loc[:, "End"])
57       last_stock = None
58       stock_df = None
59       cnt = 0
60       for rownum in range(df.shape[0]):
61           stock = df.loc[rownum, "Stock"]
62           begin = df.loc[rownum, "Begin"]
63           end = df.loc[rownum, "End"]
64           if stock != last_stock:
65               stock_df = pd.read_csv(os.path.join(price_
                 data_dir, "%s.csv" % stock))
66               stock_df.loc[:, "Date"] = pd.to_
                 datetime(stock_df.loc[:, "Date"])
67               cnt = 0
68
69           ibeg = stock_df.loc[stock_df.loc[:, "Date"].
             eq(begin), :].index[0]
70           iend = stock_df.loc[stock_df.loc[:, "Date"].
             eq(end), :].index[0]
71
72           dplt = DatePlotter()
73           plt = dplt.plot(stock_df.loc[ibeg:iend, :],
             'Date', ["Adj Close"], xlabel='Date',
```

```
74                          ylabel='Price', labels=[stock],
                            round='D')
75          plt.title("Cup-and-Handle in %s" % stock)
76          # plt.show()
77          filename = os.path.join(output_dir, "%s_%d.png" %
            (stock, cnt))
78          plt.savefig(filename)
79          cnt = cnt + 1
80
81   def buildModel():
82       model = models.Sequential()
83       model.add(layers.Conv2D(10, (3, 3),
         activation="relu", input_shape=(20, 20, 1)))
84       model.add(layers.AveragePooling2D(pool_size=(2, 2)))
85       model.add(layers.Conv2D(5, (4, 4),
         activation="relu"))
86       model.add(layers.Flatten())
87       model.add(layers.Dense(2, activation="relu"))
88       model.add(layers.Dense(2))
89       model.summary()
90       loss_fn = tf.keras.losses.SparseCategoricalCrossentro
         py(from_logits=True)
91       model.compile(optimizer='adam',
92                     loss=loss_fn,
93                     metrics=['accuracy'])
94       return model
95
96
97   def trainModel(model, df, training_rows):
98       data = np.transpose(df.reset_index(drop=True).values)
```

```
99      y_actual = np.array([int(c.startswith("t")) for c in
        df.columns], dtype=np.int)
100     train_data = data[0:training_rows, :]
101     train_data_final = np.zeros((training_rows, 20, 20,
        1), dtype=np.float32)
102     for i in range(training_rows):
103         for j in range(20):
104             pixel = int(train_data[i, j] * 20)
105             if pixel == 20:
106                 pixel = 19
107             train_data_final[i, j, pixel, 0] = 1
108     train_output = y_actual[0:training_rows]
109     model.fit(train_data_final, train_output, epochs=5)
110
111     validation_data = data[training_rows:, :]
112     validation_dt = np.zeros((validation_data.shape[0],
        20, 20, 1), dtype=np.int)
113     for i in range(training_rows, validation_
        dt.shape[0]):
114         for j in range(20):
115             pixel = int(train_data[i, j] * 20)
116             if pixel == 20:
117                 pixel = 19
118             validation_dt[i, j, pixel, 0] = 1
119     validation_output = y_actual[training_rows:]
120     model.evaluate(validation_dt, validation_output,
        verbose=2)
121     #predictions = model(x_train[:1]).numpy()
122     # this is a probabilistic model, add a softmax layer
        at the end
```

```
123     new_model = tf.keras.Sequential([model, tf.keras.
        layers.Softmax()])
124     return new_model
125
126
127 def rescaleXDimension(ar, xsize):
128     if ar.shape[0] == xsize:
129         return ar
130
131     if ar.shape[0] > xsize:
132         px = ar
133         px2 = np.zeros(xsize, dtype=np.float64)
134         px2[0] = px[0]
135         px2[-1] = px[-1]
136         delta = float(ar.shape[0])/xsize
137         for i in range(1, xsize-1):
138             k = int(i*delta)
139             fac1 = i*delta - k
140             fac2 = k + 1 - i*delta
141             px2[i] = fac1 * px[k+1] + fac2 * px[k]
142
143         return px2
144     raise ValueError("df rows are less than required
        price array elements")
145
146
147 def identify(model, df_stock, ndays, stock, res_df):
148     px_arr = df_stock.loc[:, "Adj Close"].values
149     date_arr = df_stock.loc[:, "Date"].values
150     days_identified = set(res_df.loc[res_df.loc[:,
        "Stock"].eq(stock), "Begin"])
```

```
151        inp = np.zeros((1, 20, 20, 1), dtype=np.float32)
152        for i in range(df_stock.shape[0] - ndays):
153            if date_arr[i] in days_identified:
154                continue
155            inp[:, :, :, :] = 0
156            px = px_arr[i:i+ndays]
157            mn = px.min()
158            mx = px.max()
159            transform_px = np.divide(np.subtract(px,
                   mn), mx-mn)
160            transform = rescaleXDimension(transform_px, 20)
161            for j in range(20):
162                vl = int(transform[j] * 20)
163                if vl == 20:
164                    vl = 19
165                inp[0, j, vl, 0] = 1
166
167            outval = model(inp).numpy()
168            if outval[0, 1] >= 0.9:
169                print("%s from %s - %s dates" % (stock, date_
                   arr[i], date_arr[i+ndays-1]))
170                res_df = res_df.append({"Stock":stock,
                   "Begin": date_arr[i], "End": date_
                   arr[i+ndays-1]},
171                                        ignore_index=True)
172        return res_df
173
174
175    def processData(stock_list, input_dir, price_data_dir,
       output_dir):
```

```
176        res_df = pd.DataFrame(data={"Stock":[], "Begin":[],
           "End":[]})
177        df = pd.read_csv(os.path.join(input_dir,
           "train.csv"))
178        df.drop(columns=["Day"], inplace=True)
179        obs = len(df.columns)
180        training_perc = 0.95
181        train_rows = int(obs * training_perc)
182        model = buildModel()
183        model = trainModel(model, df, training_
           rows=train_rows)
184
185        # predict
186        period_begin = 40
187        period_end = 70
188        for stock in stock_list:
189            df_stock = pd.read_csv(os.path.join(price_data_
               dir, "%s.csv"%stock))
190            for period in range(period_begin, period_end):
191                res_df = identify(model, df_stock, period,
                   stock, res_df)
192        res_df.to_csv(os.path.join(output_dir, "ch_out.csv"),
           index=False)
193
194
195    if __name__ == "__main__":
196        input_dir = r"C:\prog\cygwin\home\samit_000\value_
           momentum_new\value_momentum\data"
197        price_data_dir = r"C:\prog\cygwin\home\samit_000\
           value_momentum_new\value_momentum\data\price"
```

```
198    output_dir = r"C:\prog\cygwin\home\samit_000\value_
       momentum_new\value_momentum\output\pattern"
199    df = pd.read_table(os.path.join(input_dir, "dow.
       txt"), header=None)
200    stocks = ["TRV", "IBM"] # df.loc[:, 0].values
201    processData(stocks, input_dir, price_data_dir,
       output_dir)
202    plotData(price_data_dir, output_dir)
```

# 3.6 Using CNNs for Recognizing Handwritten Digits

Identifying handwritten digits and characters is an essential component of automated tools like mobile check deposit processors and digital assistants. Mobile check deposits are now a ubiquitous feature of most mobile banking apps for smartphones. At its heart, these tools recognize digits and characters and convert them to their digital counterparts. Recognizing digits is related to shape recognition – an objective well suited for CNNs.

In this example, let us build a CNN model for recognizing handwritten digits. The MNIST dataset of handwritten digits comprises a training set with 60,000 images and a testing set with 10,000 images. It is available from the **tf.keras.datasets.mnist** dataset. Digits can be from 0 to 9, that is, ten classes. Each image is a 28 by 28–pixel grayscale image. The digits have been centered and normalized in size to fit a 28 by 28–pixel window.

Due to the similarity of the MNIST handwritten digit dataset with the fashion MNIST dataset, all we need to do is change the code to read the MNIST dataset and change the classes, that is, change lines 15 and 16 of code in Listing 3-8 to code from Listing 3-11. During training, the CNN learns to detect images as digits from 0 to 9, without any more code changes. This illustrates the generality and elegance of CNNs in computer vision.

**Listing 3-11.** 2D Convolutional Neural Network for Identifying Handwritten Digits – Change Two Lines from Fashion MNIST Code

```
1
2   (trainx, trainy), (testx, testy) = tf.keras.datasets.mnist.
    load_data()
3   self.classes = list(range(10))
```

The CNN model finds this classification task simpler than the fashion MNIST problem. This can be seen from the plots of accuracy and loss in Figures 3-10 and 3-9. Sparse categorical accuracy begins at 95% in the first epoch on the training dataset and quickly reaches 99.7% by the tenth epoch on the training dataset. By contrast, the fashion MNIST model's accuracy begins at around 86% in the first epoch and reaches 99% by the 20th epoch. For this problem, 20 epochs are sufficient for training to converge.

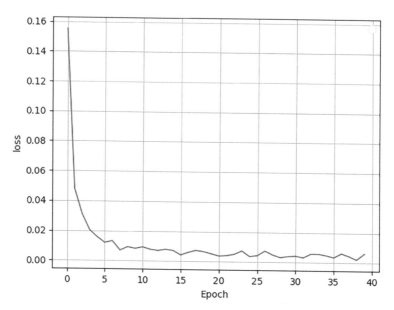

**Figure 3-9.** *Evolution of Sparse Categorical Cross Entropy Loss with Epochs Using CNN*

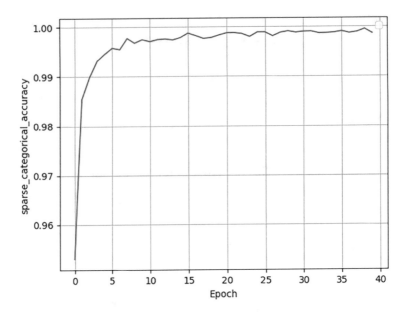

***Figure 3-10.*** *Evolution of Sparse Categorical Accuracy Metric with Epochs Using CNN*

Confusion matrices for training and testing dataset predictions are shown in Figures 3-11 and 3-12 and depict the performance of the classifier. The classifier attains 99.9% accuracy on the training dataset and 98.4% accuracy on the testing dataset.

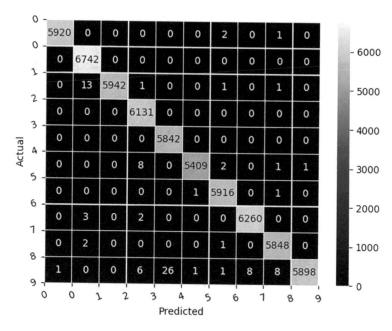

**Figure 3-11.**  *Confusion Matrix of CNN Model Predictions (X) Against Actual Labels (Y) on the Training Dataset*

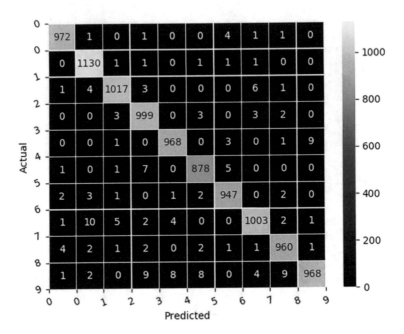

**Figure 3-12.** *Confusion Matrix of CNN Model Predictions (X) Against Actual Labels (Y) on the Testing Dataset*

# CHAPTER 4

# Recurrent Neural Networks

A recurrent neural network (RNN) is applied to inputs recurrently, with the network output from one time step sending an additional input to the next time step, augmenting the input for that time step. Inputs can be observations recorded at different time steps. Recurrent application of the network enables such networks to detect temporal relationships in input data that have a material impact in modeling output. The network's output from one time step is passed as an input to the same network at the next time step, along with inputs for that time step. This enables RNNs to pass information learned from one time step to subsequent ones. This chapter illustrates the use of recurrent neural networks by focusing on gated recurrence unit (GRU), long-short-term memory (LSTM) cell, and customized recurrent neural network layers. RNNs are trained using backpropagation through time (BPTT), which involves unrolling the network through time and using backpropagation. Vanishing gradients pose a challenge to training RNNs, as the examples will demonstrate. A LSTM network was proposed by Hochreiter and Schmidhuber in 1997. In 2007, it was applied to speech recognition with outstanding results.

TensorFlow supports three layers for building RNNs as described in the following. Each of these layers is discussed in the following sections:

© Samit Ahlawat 2023
S. Ahlawat, *Reinforcement Learning for Finance*,
https://doi.org/10.1007/978-1-4842-8835-1_4

1. **tf.keras.layers.SimpleRNN** consists of a simple
   recurrent neural network cell that accepts output
   from the previous time step's simple RNN cell,
   a bias, and inputs from the current time step to
   generate an output. Let **f** denote the cell's activation
   function, $X_t$ denote the input vector of length **n** at
   time t, and b denote bias. The cell's output is shown
   in equation 4.1. This cell has **n + 2** free parameters.

$$y_t = f\left( \sum_i W_i X_{t,i} + W_b b + W_c y_{t-1} \right)$$

(4.1)

2. **tf.keras.layers.LSTM** comprises of long-short-
   term memory cells. These cells have the additional
   capability of forgetting previous cell outputs.

3. **tf.keras.layers.GRU** consists of a simple
   gated recurrence unit cell as compared with
   LSTM. However, it has all the essential features
   of LSTM.

4. **tf.keras.layers.RNN** layer is useful for the definition
   of customized RNN layers.

# 4.1 Simple RNN Layer

In its simplest form, an RNN consists of a neuron applied recurrently to
inputs. This is illustrated in Figure 4-1. In TensorFlow, this can be written
as shown in Listing 4-1 using **SimpleRNN**. TensorFlow's **SimpleRNN** layer
consists of a set of neurons applied recurrently to input. It has hyperbolic
tangent as the default activation function. By default, **SimpleRNN** returns
the final output corresponding to the last time step as the network

output. This behavior can be changed by using the argument **return_ sequences=True**, so that it returns the cell output from each time step. Model summary is shown in Listing 4-2. It is instructive to study the shapes of input and output vectors in order to understand the SimpleRNN layer. Input shape is (None, 4, 4), representing four time steps each with four inputs. According to TensorFlow convention, the first dimension of input represents the number of batches. Trainable parameters are six, four corresponding to input weights and one each for bias and the previous time step's cell output.

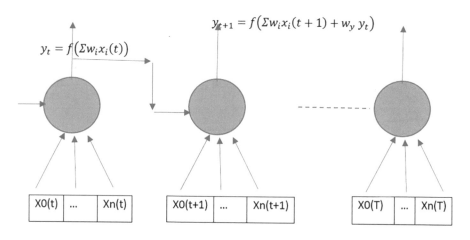

$$y_t = f\left(\Sigma w_i x_i(t)\right)$$

$$y_{t+1} = f\left(\Sigma w_i x_i(t+1) + w_y\, y_t\right)$$

Recurrent Neural Network with a Single Cell, Applying Activation f(.)

***Figure 4-1.*** *Simple RNN Cell*

***Listing 4-1.*** Example of a Recurrent Neural Network with One Cell

```
1    import tensorflow as tf
2    from tensorflow.keras import layers, models
3
4    model = models.Sequential()
5    lyr = tf.keras.layers.SimpleRNN(1, return_sequences=True)
6    model.add(lyr)
```

```
7    input_shape = (None, 4, 4)
8    model.build(input_shape)
9    print(model.summary())
10   model.compile(optimizer="adam", loss=tf.keras.losses.
     MeanSquaredError(), metrics=["mse"])
```

***Listing 4-2.*** Model Summary for a Network with the SimpleRNN Layer

```
1    Model: "sequential"
2    _____
3    Layer (type)               Output Shape              Param #
4    _____
5    simple_rnn (SimpleRNN)     (None, 4, 1)              6
6
7    _____
8    Total params: 6
9    Trainable params: 6
10   Non-trainable params: 0
11   _____
```

The example shown in Listing 4-3 shows a sequential layer built with the SimpleRNN layer. The **input_shape** argument of the **SimpleRNN** layer is a tuple comprising of the number of recurrent steps the layer is applied and the number of features accepted by the **SimpleRNN** cell. The number of recurrent steps is different from the number of batches, which can be left as **None**. Actual input to this layer is three-dimensional: number of batches, number of recurrent time steps, and number of features, which is (2, 4, 5) in Listing 4-3. The number of trainable parameters for the **SimpleRNN** layer is the sum of the number of features, number of cells, and a bias multiplied by the number of cells, as shown in equation 4.2. The simple RNN cell sends a cell state from one time step to the next, and there are $N_{cells}$ number of cells, giving $N_{cells}$ outputs from one time step sent to the next one. There is one bias term. For the example in Listing 4-3, the **SimpleRNN** layer has 160 trainable parameters $(5 + 10 + 1)10$.

$$N_{SimpleRNN} = \left(N_{features} + N_{cells} + 1\right)N_{cells} \qquad (4.2)$$

***Listing 4-3.*** Simple RNN Layer Inside a Sequential Model

```
1    import tensorflow as tf
2    model = tf.keras.Sequential()
3    model.add(tf.keras.layers.SimpleRNN(10, input_
     shape=(None, 5)))
4    model.add(tf.keras.layers.Dense(6))
5    print(model.summary())
6
7    input = tf.constant(tf.ones((2, 4, 5)))
8    output = model(input)
9    print(output.shape)
10
11   _____
12   Layer (type)              Output Shape                Param #
13   _____
14   simple_rnn_1 (SimpleRNN)  (None, 10)                     160
15
16   dense_1 (Dense)           (None, 6)                       66
17
18   _____
19   Total params: 226
20
21   (2, 6)
```

Output of the **SimpleRNN** layer has shape $(N_{batches}, N_{cells})$ if **return_sequences** is set to **False** and has shape $(N_{batches}, N_{steps}, N_{cells})$ otherwise.

## 4.2 LSTM Layer

A LSTM (long-short-term memory) cell is more complex than a simple RNN cell. LSTM transmits cell state in addition to cell output to the next time step. It has four internal gates to control the flow of inputs: forget gate to control transmission of cell state from the last time step, update gate to control the update to cell state, tanh gate that is used along with the update gate, and output gate that controls the output of the cell. Like all RNNs, LSTM is applied recurrently to inputs from successive time steps. A LSTM cell is shown in Figure 4-2. The flow of information along with transformations applied in LSTM is described in the following:

1.  Cell state $C_{t-1}$ and cell activation $a_{t-1}$ from time step $t-1$ flow as inputs to the cell at time $t$.

2.  Cell output is denoted by $y_t$ and is only returned for the cell corresponding to final time step $T$ by default. Outputs from all time steps can be returned by passing the argument **return_sequences=True**.

3.  Input for time step $t$ is denoted by $X_t$. This could be a vector. As an example, the simple RNN cell in Listing 4-1 has four components in the input at each time step.

4.  The forget gate is applied to input vector $X_t$ and cell activation from the last time step, $a_{t-1}$. It uses a sigmoid function as the default activation function, as shown in equation 4.3:

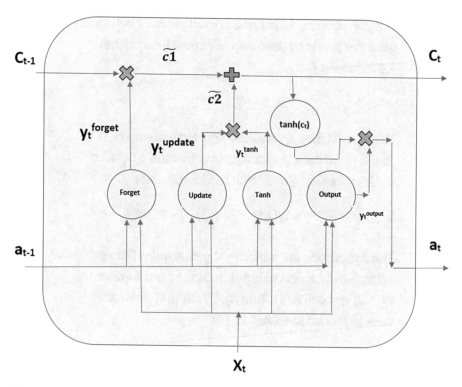

***Figure 4-2.*** *LSTM Cell*

$$y_t^{forget} = \sigma\left(W_x^f \cdot X_t + W_a^f a_{t-1} + W_b^f b^f\right)$$
$$\sigma(x) = \frac{1}{1+e^{-x}} \tag{4.3}$$

This gate has $N + 2$ free parameters, where $N$ denotes the dimension of $X_t$. Output from this gate is multiplied by cell state from the previous time step. This gate is called the forget gate because if $y_t^{forget}$ is 0, cell state from the previous step will have no impact on the output and will in effect be forgotten.

5. Output from the forget gate is multiplied by cell state from the previous time step as shown in equation 4.4 to obtain $\tilde{c}_1$:

$$\tilde{c}_1 = C_{t-1} y_t^{forget} \tag{4.4}$$

6. The update gate applies the sigmoid activation function to input vector $X_t$ and activation $a_{t-1}$ from the last time step, as shown in equation 4.5:

$$y_t^{update} = \sigma\left(W_x^u \cdot X_t + W_a^u a_{t-1} + W_b^u b^u\right) \tag{4.5}$$

7. The hyperbolic tangent (tanh) gate applies the tanh activation function to input from the current time step $X_t$ and cell activation $a_{t-1}$ from the last time step, as shown in equation 4.6:

$$y_t^{tanh} = tanh\left(W_x^{tanh} \cdot X_t + W_a^{tanh} a_{t-1} + W_b^{tanh} b^t\right)$$
$$tanh(x) = \frac{e^x + e^{-x}}{e^x - e^{-x}} \tag{4.6}$$

8. Output from the update gate and tanh gate is multiplied to get $\tilde{c}_2$ as shown in equation 4.7:

$$\tilde{c}_2 = y_t^{tanh} y_t^{update} \tag{4.7}$$

9. Cell state output is obtained by adding $\tilde{c}_1$ and $\tilde{c}_2$ as shown in equation 4.8. This cell state is sent to the next time step:

$$C_t = \tilde{c}_1 + \tilde{c}_2 \tag{4.8}$$

10.  The output gate applies sigmoid activation to input $X_t$ and cell activation from the previous time step $a_{t-1}$, as shown in equation 4.9:

$$y_t^{output} = \sigma\left(W_x^o \cdot X_t + W_a^o a_{t-1} + W_b^o b^o\right) \tag{4.9}$$

11.  Cell activation $a_t$ is calculated using equation 4.10:

$$a_t = tanh\left(C_t\right) y_t^{output} \tag{4.10}$$

12.  Cell output is calculated by applying an activation function to $a_t$. If no activation function is specified, TensorFlow takes it to be a unit transformation, that is, it returns the same input as output, $y_t = a_t$. If an activation function $f$ is specified, cell output would be $y_t = f(a_t)$.

The number of trainable parameters of a LSTM cell is given by equation 4-11. 4 corresponds to the number of gates in a LSTM cell. Output of the LSTM layer has shape ($N_{batch}$, $N_{steps}$, $N_{cells}$) when **return_sequences** is set to **True** and shape ($N_{batch}$, $N_{cells}$) when it is **False**.

$$N_{LSTM} = 4\left(N_{features} + N_{cells} + 1\right)N_{cells} \tag{4.11}$$

Example of a LSTM layer in a neural network is shown in Listing 4-4.

**Listing 4-4.**  LSTM Layer Inside a Sequential Model

```
1   import tensorflow as tf
2   model = tf.keras.Sequential()
3   model.add(tf.keras.layers.LSTM(10, input_shape=(None, 5),
    return_sequences=True))
4   model.add(tf.keras.layers.Dense(6))
5   print(model.summary())
6
```

```
7    Model: "sequential_2"
8    _____
9    Layer (type)              Output Shape              Param #
10   _____
11   lstm (LSTM)               (None, None, 10)          640
12
13   dense_2 (Dense)           (None, None, 6)           66
14
15   _____
16   Total params: 706
17   Trainable params: 706
18   Non-trainable params: 0
19   _____
20
21
22   input = tf.constant(tf.ones((2, 4, 5)))
23   output = model(input)
24   print(output.shape)
25
26   (2, 4, 6)
```

# 4.3 GRU Layer

Gated recurrent unit (GRU) is a simplified version of LSTM that sends only one output $h_t$ to the next time step and has three gates – reset gate, update gate, and activation gate. Recall that a LSTM cell sends output $y_t$ and cell state $C_t$ to the cell at the next time step and has four gates.

Output produced by GRU can be understood by looking at output of cell gates. At time step $t$, a GRU cell receives output $h_{t-1}$ from GRU at the previous time step and input $X_t$.

1.  The **update gate** uses previous time step output
    $h_{t-1}$ and input $X_t$ to compute activation, as shown in
    equation 4.12. The activation function is sigmoid by
    default:

$$z_t = \sigma\left(W_x^u \cdot X_t + W_h^u h_{t-1} + W_b^u b_u\right)$$

$$\sigma(y) = \frac{1}{1+e^{-y}} \tag{4.12}$$

2.  The **reset gate** similarly applies the activation
    function to a dot product of weights and $h_{t-1}$ and $X_t$
    as shown in equation 4.13:

$$r_t = \sigma\left(W_x^r \cdot X_t + W_h^r h_{t-1} + W_b^r b_r\right) \tag{4.13}$$

3.  The **activation gate** takes a product of output from
    the reset gate and previous time step output $h_{t-1}$
    along with a dot product of weights and input vector
    $X_t$ to get activation $\hat{h}_t$ after applying the hyperbolic
    tangent activation function as shown in equation
    4.14. The role of the reset gate is illustrated in
    equation 4.14. If reset gate output $r_t$ is zero, output
    from the last time step's GRU cell is ignored:

$$\hat{h}_t = tanh\left(W_x^h \cdot X_t + W_r^h r_t \cdot h_{t-1} + W_b^h b^t\right)$$

$$tanh(x) = \frac{e^x + e^{-x}}{e^x - e^{-x}} \tag{4.14}$$

4. The final cell state $h_t$ is calculated by interpolating between previous time step's cell state $h_{t-1}$ and this cell's output $\hat{h}_t$ using output $z_t$ from the output gate as interpolation factor, as shown in equation 4.15:

$$h_t = z_t \hat{h}_t + (1 - z_t) h_{t-1} \qquad (4.15)$$

GRU was first introduced by Cho et al. in 2014 in an application of an RNN encoder-decoder model applied to language translation.

# 4.4 Customized RNN Layers

Customized RNN layers can be created in TensorFlow by first defining a customized RNN cell and passing it to the constructor of class **tf.keras. layers.RNN**. Let us create an RNN cell that takes the cell outputs from previous two time steps as input, in addition to the input features from the current time step, to produce an output. The code for this cell is shown in Listing 4-5. Due to random weight initializers, actual output may differ from the one shown in Listing 4-5.

***Listing 4-5.*** Using an RNN Layer to Create Customized RNN Layers

```
1    import tensorflow as tf
2
3    class CustomRNN(tf.keras.layers.Layer):
4        def __init__(self, units, **kwargs):
5            self.nunit = units
6            self.state_size = units
7            self.prev2Output = None
8            super().__init__(**kwargs)
9
10       def build(self, input_shape):
```

```
11            self.xWt = self.add_weight(shape=(input_
              shape[-1], self.nunit),
12                                      initializer=tf.
                                        keras.initializers.
                                        RandomNormal(),
13                                      name="xWt")
14            self.h1Wt = self.add_weight(shape=(self.nunit,
              self.nunit),
15                                      initializer=tf.
                                        keras.initializers.
                                        RandomNormal(),
16                                      name="h1")
17            self.h2Wt = self.add_weight(shape=(self.nunit,
              self.nunit),
18                                      initializer=tf.
                                        keras.initializers.
                                        RandomNormal(),
19                                      name="h2")
20            self.built = True
21
22        def call(self, inputs, states):
23            prevOutput = states[0]
24            output = tf.matmul(inputs, self.xWt) +
              tf.matmul(prevOutput, self.h1Wt)
25            if self.prev2Output is not None:
26                output += tf.matmul(self.prev2Output,
                    self.h2Wt)
27            self.prev2Output = prevOutput
28            return output, [output]
29
30    cell = CustomRNN(5)
```

```
31    layer = tf.keras.layers.RNN(cell)
32    input = tf.ones((2, 6, 5))
33    y = layer(input)
34    print(y)
35    print(y.shape)
36
37    <tf.Tensor: shape=(2, 5), dtype=float32, numpy=
38    array([[-0.13667805,  0.11874562, -0.03024731,
      -0.04962897,  0.0992294 ],
39    [-0.13667805,  0.11874562, -0.03024731,
      -0.04962897,  0.0992294 ]],
40    dtype=float32)>
41
42    (2, 5)
```

# 4.5 Stock Price Prediction

Stock price prediction is a cornerstone financial modeling problem that has drawn keen research interest over decades. The problem involves predicting stock price at future time intervals given a history of predictor variables. Researchers have used a variety of predictor variables in myriad modeling methodologies to predict stock price. For example, Campbell and Schiller (1988) used dividend yield to predict stock returns, Lakonishok et al. (1994) investigated the predictive power of value measures such as price-to-earnings and book-to-market value in stock price prediction, Chan et al. (1996) applied momentum measure of stock price return to predict future returns, and Fama and French (2015) applied a five-factor model that includes market return, return on small market capitalization minus return on big market capitalization stocks, return on high minus low book-to-market value stocks, return on high minus low investment firms, and return on high-profitability stocks minus return on low-profitability stocks. On the methodology side, there is an equally diverse spectrum of models applied to this problem: from simple linear regression used by Fama and French (2015) and simple technical trading rules (Brock et al., 1992) to genetic algorithms (Allen et al., 1999) and probabilistic neural networks (Ahlawat, 2016).

Stock price returns display varying degrees of autocorrelation. Intuitively, one would expect a stock that has positive return over 1 day to have positive return the next day. Many stocks, including the S&P 500 index, for example, have a high degree of mean reversion, meaning that a high positive return on a day is followed by a negative return the following day. While daily returns have more volatility, monthly returns have less volatility, implying higher predictability. Because recurrent neural networks transmit information from one period to the next, they are a natural tool to employ for capturing stock price return autocorrelation. This section applies RNNs to predict monthly price return of the S&P 500 index.

S&P 500 is an index comprising of 500 publicly traded large-capitalization stocks in the United States. It is one of the most widely tracked market indices, serving as a gauge for market performance. In this section, let us use an RNN to predict 1-month return on the S&P 500 index. Data consists of daily closing price and traded share volume of SPY – an S&P 500 tracking ETF – from January 2000 to July 2022. A recurrent neural network is built to predict 1-month return of SPY, and its prediction accuracy is compared against a baseline predictor that uses last month's return as 1-month return prediction. RNN layers **SimpleRNN**, **GRU**, and **LSTM** are compared with each other to see which layer gives better prediction accuracy in training and testing datasets. Feature selection, considerations for model building, and results are discussed in the following.

1.  Four features are used in the model. Data is available for trading days only. One month is defined as 21 trading days because a month has 21 trading days on average. Likewise, a year is defined as 252 trading days:

    •   Last 1-month return $r_t$ calculated as $\dfrac{P_t - P_{t-21\,days}}{P_{t-21\,days}}$.

        A month has 21 trading days on average.

    •   Momentum factor $m_t$ that represents the price momentum. It is calculated using prior 1-month return $r_t$ and prior 3-month return $\tilde{r}_t$ as shown in equation 4-16. Three months equate to 63 trading days on average:

    $$\tilde{r}_t = \frac{P_t - P_{t-63\,days}}{P_{t-63\,days}}$$

    $$m_t = \frac{r_t}{|\tilde{r}_t| + |r_t}$$

    (4.16)

    •   Volatility factor $v_t$ that describes the extent of volatility observed in price returns. It is defined as the ratio of variance in price return observed over the last 1 month (21 days) and the average 1-month variance of returns observed over the last year (or 252 trading days), as shown in equation 4.17:

$$\mu_t \quad = \frac{\sum_{i=1}^{2} 1 r_{t-i}}{21}$$

$$\sigma_t^2 \quad = \frac{\sum_{i=1}^{2} 1 \left(r_{t-i} - \mu_t\right)^2}{21}$$

$$\mu\left(\sigma_t^2\right) \quad = \frac{\sum_{i=1}^{2} 52 \sigma_{t-i}^2}{252}$$

$$\nu_t \quad = \frac{\sigma_t^2}{\mu\left(\sigma_t^2\right)}$$

(4.17)

- Volume factor $\nu_t$ defined as the ratio of traded shares on a day to the average volume of traded shares over the last month (last 21 trading days), as shown in equation 4.18:

$$\mu\left(Volume_t\right) = \frac{\sum_{i=1}^{2} 1 Volume_{t-i}}{21}$$

$$V_t = \frac{Volume_t}{\mu\left(Volume_t\right)}$$

(4.18)

2. A boxplot of features is shown in Figure 4-3. As seen in the figure, all input features are in the range of around 5 to –1. This means that feature normalization is not required.

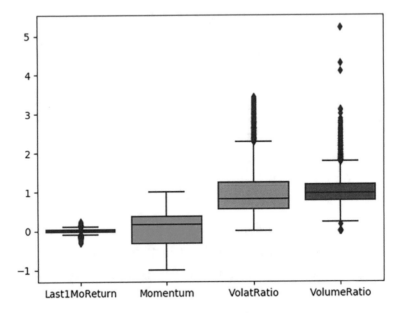

***Figure 4-3.*** *Boxplot of Input Features in the Training Dataset*

3.  A train-test split of 70%-30% is used.

4.  The number of recurrent timesteps $N_t$ is selected
    as 10 trading days. State information flows only in
    $N_t$ days. This means that the RNN can only identify
    autocorrelations and other temporal relationships
    over $N_t$ or 10 trading days. Increase $N_t$ to enable
    the RNN to identify temporal relationships over a
    longer period.

5.  Input data for the RNN model is converted to a
    three-dimensional matrix of dimensions $(N_{batches}, N_t, N_{features})$. In this example, the number of features
    $N_{features}$ is 4.

6.  The mean square error loss function is used.

7.  A plot of loss history (Figure 4-4) shows that training converges after about 30 epochs.

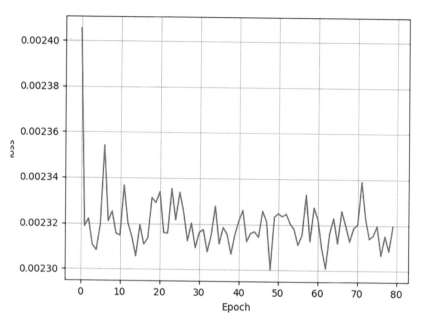

**Figure 4-4.** *History of Mean Square Error Showing Convergence*

8.  To compare the performance of RNN models, a baseline model that predicts 1-month return as the last 1-month return is added.

9.  Results show that LSTM performs the best in training and testing datasets, followed by GRU and SimpleRNN layers. All three RNN models perform significantly better than the baseline model. Loss function values for the four models on training and testing datasets are shown in Table 4-1.

***Table 4-1.*** *Comparison of Standard Deviation Between Predicted and Actual Returns*

| Dataset | LSTM | GRU | SimpleRNN | Baseline |
|---------|------|-----|-----------|----------|
| Training | 0.002296 | 0.002345 | 0.002312 | 0.004782 |
| Testing | 0.002233 | 0.002255 | 0.002873 | 0.004938 |

10. Using predicted return values, predicted stock price is calculated and plotted against actual stock price. Using the LSTM layer, predicted vs. actual stock price of SPY is shown in Figure 4-5 for training data and in Figure 4-6 for testing data. As can be seen, the fit is generally good, except when prices witness a steep decline. Predicted vs. actual price plots for the last 2 years in training and testing datasets illustrate this point more clearly, as seen in Figures 4-7 and 4-8.

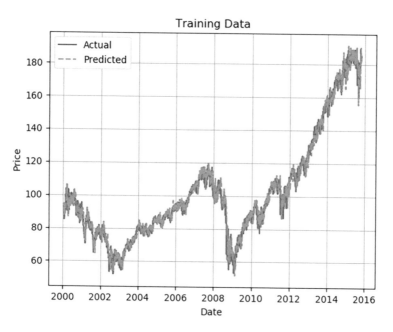

**Figure 4-5.** *Predicted vs. Actual SPY Price in the Training Dataset*

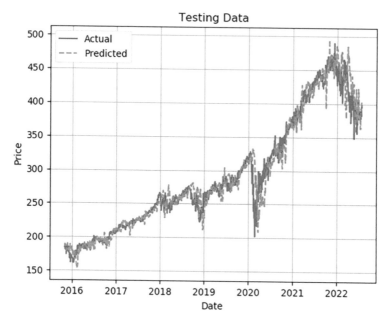

**Figure 4-6.** *Predicted vs. Actual SPY Price in the Testing Dataset*

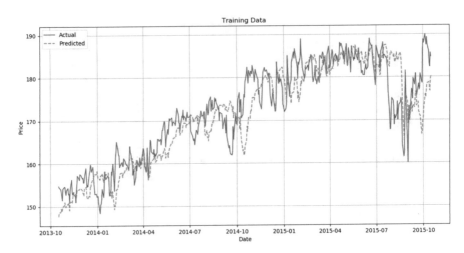

***Figure 4-7.*** *Predicted vs. Actual SPY Price for the Last 2 Years in the Training Dataset*

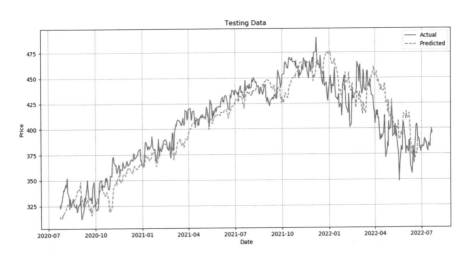

***Figure 4-8.*** *Predicted vs. Actual SPY Price for the Last 2 Years in the Testing Dataset*

The complete code for this example is shown in Listing 4-6.

***Listing 4-6.*** Predict the S&P 500 Tracking ETF's Price Using RNN

```
1    import numpy as np
2    import pandas as pd
3    import tensorflow as tf
4    import matplotlib.pyplot as plt
5    import seaborn as sns
6    import os
7    import logging
8
9    logging.basicConfig(level=logging.DEBUG)
10
11
12   class ReturnPredictor(object):
13       def __init__(self, dirname, trainTestSplit=0.7,
         nunit=15, ntimestep=10, batchSize=10, nepoch=40):
14           filename = os.path.join(dirname, "SPY.csv")
15           self.inputDir = dirname
16           self.logger = logging.getLogger(self.__
             class__.__name__)
17           df = pd.read_csv(filename)
18           self.nUnit = nunit
19           self.nTimestep = ntimestep
20           self.dateCol = "Date"
21           self.priceCol = "Adj Close"
22           self.volCol = "Volume"
23           self.volatilityCol = "VolatRatio"
24           self.volumeCol = "VolumeRatio"
25           self.momentumCol = "Momentum"
26           self.returnCol = "Last1MoReturn"
```

```
27          self.resultCol = "Fwd1MoReturn"
28          self.daysInMonth = 21
29          df = self.featureEngineer(df)
30          ntrain = int(trainTestSplit * df.shape[0])
31          self.trainDf = df.loc[14*self.
            daysInMonth+1:ntrain, :].reset_index(drop=True)
32          self.testDf = df.loc[ntrain:, :].reset_
            index(drop=True)
33          self.featureCols = [self.returnCol, self.
            momentumCol, self.volatilityCol, self.volumeCol]
34          #self.plotData(self.trainDf)
35          #self.plotData(self.testDf)
36          self.nnet = self.model()
37          self.rnnTrainData = self.prepareDataForRNN(self.
            trainDf)
38          self.rnnTestData = self.prepareDataForRNN
            (self.testDf)
39          self.batchSize = batchSize
40          self.nEpoch = nepoch
41          self.cellType = None
42
43      def featureEngineer(self, df: pd.DataFrame) ->
        pd.DataFrame:
44          df.loc[:, self.dateCol] = pd.to_datetime(df.
            loc[:, self.dateCol])
45          # 1 Month lagged returns
46          returns = np.zeros(df.shape[0], dtype=np.
            float32)
47          nrow = df.shape[0]
48          returns[self.daysInMonth+1:] = np.divide(df.
            loc[self.daysInMonth:nrow-2, self.
            priceCol].values,
```

```
49                                      df.loc[0:nrow-self.
                                        daysInMonth-2, self.
                                        priceCol].values) - 1
50          df.loc[:, self.returnCol] = returns
51          # momentum factor
52          momentum = np.zeros(df.shape[0], dtype=np.
            float32)
53          returns3Mo = np.divide(df.loc[3*self.
            daysInMonth:nrow-2, self.priceCol].values,
54                                  df.loc[0:nrow-3*self.
                                        daysInMonth-2, self.
                                        priceCol].values) - 1
55          num = returns[3*self.daysInMonth+1:]
56          momentum[3*self.daysInMonth+1:] = np.divide(num,
            np.abs(num) + np.abs(returns3Mo))
57          df.loc[:, self.momentumCol] = momentum
58
59          # volatility factor
60          df.loc[:, self.volatilityCol] = 0
61          volatility = np.zeros(nrow, dtype=np.float32)
62          rtns = returns[self.daysInMonth+1:2*self.
            daysInMonth+1]
63          sumval = np.sum(rtns)
64          sumsq = np.sum(rtns * rtns)
65          for i in range(2*self.daysInMonth+1, nrow):
66              mean = sumval / self.daysInMonth
67              volatility[i] = np.sqrt(sumsq / self.
                daysInMonth - mean*mean)
68              sumval += returns[i] - returns[i-self.
                daysInMonth]
```

```
69          sumsq += returns[i] * returns[i] -
            returns[i-self.daysInMonth] * returns[i-
            self.daysInMonth]
70          oneyr = 12 * self.daysInMonth
71          df.loc[:, self.volatilityCol] = 0.0
72          for i in range(oneyr+2*self.
            daysInMonth+1, nrow):
73              df.loc[i, self.volatilityCol] =
                volatility[i] / np.mean(volatility[i-
                oneyr:i])
74
75          # volume factor
76          df.loc[:, self.volumeCol] = 0
77          volume = df.loc[:, self.volCol].values
78          for i in range(self.daysInMonth, nrow-1):
79              df.loc[i+1, self.volumeCol] = volume[i] /
                np.mean(volume[i-self.daysInMonth:i])
80
81          # result column
82          df.loc[:, self.resultCol] = 0.0
83          df.loc[0:nrow-self.daysInMonth-1, self.
            resultCol] = df.loc[self.daysInMonth:, self.
            returnCol].values
84          return df
85
86      def prepareDataForRNN(self, df):
87          nfeat = len(self.featureCols)
88          data = np.zeros((df.shape[0]-self.nTimestep,
            self.nTimestep, nfeat), dtype=np.float32)
89          results = np.zeros((df.shape[0]-self.nTimestep,
            self.nTimestep), dtype=np.float32)
90          raw_data = df[self.featureCols].values
```

```
91          raw_results = df.loc[:, self.resultCol].values
92          for i in range(0, data.shape[0]):
93              data[i, :, :] = raw_data[i:i+self.
                nTimestep, :]
94              results[i, :] = raw_results[i:i+self.
                nTimestep]
95          return data, results
96
97      def plotData(self, df: pd.DataFrame) -> None:
98          df = df.set_index(keys=[self.dateCol])
99          fig, axs = plt.subplots(nrows=len(self.
                featureCols)+1, ncols=1, figsize=(12, 16))
100         axs[0].plot(df.index.values, df.loc[:, self.
                priceCol].values)
101         axs[0].set_ylabel("Price")
102         for i, col in enumerate(self.featureCols):
103             axs[i+1].plot(df.index.values, df.loc[:,
                col].values)
104             axs[i+1].set_ylabel(col)
105         plt.show()
106
107         boxplot = df[self.featureCols]
108         sns.boxplot(data=boxplot)
109         plt.show()
110
111     def checkpointModel(self, nnet):
112         checkpointFile = os.path.join(self.inputDir,
                "checkpoint_spricernn_%s_wt" % self.cellType)
113         if not os.path.exists(checkpointFile):
114             nnet.predict(np.ones((20, self.nTimestep,
                len(self.featureCols)), dtype=np.float32))
```

```
115                    tf.keras.models.save_model(nnet,
                       checkpointFile, overwrite=False)
116            else:
117                nnet = tf.keras.models.load_
                   model(checkpointFile)
118            return nnet
119
120        def model(self):
121            nnet = tf.keras.Sequential()
122            nfeat = len(self.featureCols)
123            self.cellType = "LSTM"
124            nnet.add(tf.keras.layers.LSTM(self.nUnit, input_
                   shape=(None, nfeat)))
125            #nnet.add(tf.keras.layers.GRU(self.nUnit, input_
                   shape=(None, nfeat)))
126            #nnet.add(tf.keras.layers.SimpleRNN(self.nUnit,
                   input_shape=(None, nfeat)))
127            nnet.add(tf.keras.layers.Dense(5,
                   activation="relu"))
128            nnet.add(tf.keras.layers.Dense(1))
129
130            self.loss = tf.keras.losses.MeanSquaredError()
131            self.optimizer = tf.keras.optimizers.
                   Adam(learning_rate=0.005)
132            nnet.compile(optimizer=self.optimizer,
133                            loss=self.loss)
134            nnet = self.checkpointModel(nnet)
135            return nnet
136
137        def plotConvergenceHistory(self, history,
               metricName):
```

```
138        plt.plot(history.epoch, history.
           history[metricName])
139        plt.xlabel("Epoch")
140        plt.ylabel(metricName)
141        plt.grid(True)
142        #plt.legend()
143        plt.show()
144
145    def trainModel(self):
146        history = self.nnet.fit(self.rnnTrainData[0],
           self.rnnTrainData[1],
147                                batch_size=self.
                                batchSize,
                                epochs=self.nEpoch)
148        self.plotConvergenceHistory(history, "loss")
149        return history
150
151    def testModel(self):
152        mse = tf.keras.losses.MeanSquaredError()
153        cnt = 0
154        for X, y in [self.rnnTrainData, self.
           rnnTestData]:
155            predict = self.nnet.predict(X)
156            loss = mse(y[:, -1], predict[:, 0]).numpy()
157            self.logger.info("final loss = %f", loss)
158            # baseline model prediction that uses
               last month's return as prediction for 1
               month return
159            loss = mse(y[:, -1], X[:, -1, 0]).numpy()
160            self.logger.info("baseline loss = %f", loss)
161            # plot predicted vs actual vs baseline
```

```
162                self.plotPredictedReturn(y, predict[:, 0],
                   cnt == 0)
163                cnt += 1
164
165        def plotPredictedReturn(self, yActual: np.ndarray,
           yPred: np.ndarray, isTrain: bool) -> None:
166            pxActual = np.zeros(yActual.shape[0], dtype=np.
               float32)
167            pxPred = np.zeros(yActual.shape[0], dtype=np.
               float32)
168
169            dts = [None] * yActual.shape[0]
170            df = self.trainDf
171            if not isTrain:
172                df = self.testDf
173
174            for i in range(pxActual.shape[0]):
175                px = df.loc[i+self.nTimestep, self.priceCol]
176                pxActual[i] = px*(1.0 + yActual[i, -1])
177                pxPred[i] = px*(1.0 + yPred[i])
178                dts[i] = df.loc[i+self.nTimestep, self.
                   dateCol]
179            plt.plot(dts, pxActual, label="Actual")
180            plt.plot(dts, pxPred, "--", label="Predicted")
181            plt.xlabel("Date")
182            plt.ylabel("Price")
183            plt.grid(True)
184            title = "Training Data" if isTrain else
               "Testing Data"
185            plt.title(title)
186            plt.legend()
187            plt.show()
```

```
188
189             plt.plot(dts[-252*2:], pxActual[-252*2:],
                label="Actual")
190             plt.plot(dts[-252*2:], pxPred[-252*2:], "--",
                label="Predicted")
191             plt.xlabel("Date")
192             plt.ylabel("Price")
193             plt.grid(True)
194             title = "Training Data" if isTrain else
                "Testing Data"
195             plt.title(title)
196             plt.legend()
197             plt.show()
198
199
200     if __name__ == "__main__":
201         sp500file = r"C:\prog\cygwin\home\samit_000\RLPy\
            data\book"
202         rpred = ReturnPredictor(sp500file, nepoch=80)
203         rpred.trainModel()
204         rpred.testModel()
```

# 4.6  Correlation in Asset Returns

Let us use LSTM cells to identify correlation in asset returns. S&P 500 stocks have been divided into 11 diversified sectors. It is well known that some of these sectors have high correlation with market movements (e.g., financials), while other sectors that are considered conservative have lower correlations (such as utility). In this section, let us build a time-series model to predict sector returns and compare it with a neural network that has a LSTM layer.

Let us build a model to predict sector returns that depends on concurrent period market returns and lagged sector returns. Concurrent period market returns are predicted using an autoregressive model that depends on lagged market returns, in addition to the last period's market volatility and volume. The models are described briefly in the following.

First, let us build a model to predict market (S&P 500 index) weekly returns. Autocorrelation plots of weekly returns (Figure 4-9) show that taking the first five lagged returns would be sufficient. In addition, market volume observed over the past week (5 days) divided by the average volume observed during the training period and market return volatility observed over the last week are used as independent variables in the linear regression model. It can be verified that all independent variables are stationary. The fitted market model is shown in equation 4.19. Data from 2000–2015 is used for fitting the model. Weekly S&P 500 returns are negatively correlated with last week's returns, known as mean reversion.

$$
\begin{aligned}
r_M(t) \ = \ & 5.7179\times10^{-4} - 4.762\times10^{-2}\, r_M(t-1) + 3.36\times10^{-2}\, r_M(t-2) \\
& -1.891\times10^{-2}\, r_M(t-3) - 5.867\times10^{-2}\, r_M(t-4) \\
& -8.538\times10^{-3}\, r_M(t-5) - 5.05\times10^{-2}\, MVol(t-1) \\
& +1.193\times10^{-4}\, MVol(t-1) + \epsilon_M \\
\epsilon_M \ \sim \ & N\!\left(0,\sigma_M^2\right)
\end{aligned}
\tag{4.19}
$$

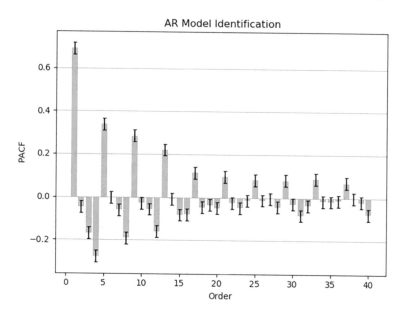

**Figure 4-9.** *Partial Autocorrelation Function of Market Returns*

The sector return model is built by regressing 1-week sector returns on concurrent market returns and lagged 1-week sector returns. The fitted coefficients are shown in Table 4-2 and equation 4.20. As can be seen, utilities have a smaller coefficient for market return than financials.

**Table 4-2.** *Sector Return Linear Regression Model Coefficients*

| Name | Sector | const | MktReturn | LaggedReturn |
|---|---|---|---|---|
| Communication Services | XLC | 0.000050 | 0.909338 | −0.027703 |
| Consumer Discretionary | XLY | 0.000716 | 1.064655 | −0.009689 |
| Consumer Staples | XLP | 0.000769 | 0.530273 | 0.000682 |
| Energy | XLE | −0.000030 | 1.091086 | 0.028018 |
| Financials | XLF | −0.000083 | 1.318176 | 0.012530 |
| Healthcare | XLV | 0.000740 | 0.778489 | −0.038654 |
| Industrials | XLI | 0.000303 | 1.080460 | 0.010503 |
| Information Technology | XLK | 0.000027 | 1.144972 | −0.014580 |
| Materials | XLB | 0.000359 | 1.083251 | −0.015964 |
| Real Estate | XLRE | −0.000463 | 0.906196 | 0.002108 |
| Utilities | XLU | 0.000785 | 0.643267 | −0.001537 |

$$r_{\text{sec}}(t) = \alpha + \beta r_M(t) + \gamma r_{\text{sec}}(t-1) + \epsilon_S \epsilon_S \sim N\left(0, \sigma_S^2\right) \qquad (4.20)$$

Next, we build a neural network model using a LSTM layer. The model definition is shown in Listing 4-7. An eight-cell LSTM layer is used in the network, followed by two dense layers. The model is trained on data from 2000 to 2015, just like the linear regression model.

**Listing 4-7.** LSTM Model for Predicting Sector Returns

```
1    import numpy as np
2    import pandas as pd
3    import os.path
4    import statsmodels.tsa.stattools
5    import matplotlib.pyplot as plt
```

```
6      import statsmodels.api as sm
7      import statsmodels.regression.linear_model as lm
8      import tensorflow as tf
9      from tensorflow.keras import layers, models
10     import itertools
11     from matplotlib.dates import DateFormatter, YearLocator,
       MonthLocator
12
13
14     class ColumnConfig(object):
15         def __init__(self):
16             self.CLOSE_PRICE = 'Adj Close'
17             self.VOLUME = 'Volume'
18             self.DATE = 'Date'
19             # Date is index
20
21
22     class TransformedRiskMeasure(object):
23         def __init__(self, name):
24             self.name = name
25
26         def calculateEMA(self, data_arr, ema=10):
27             ema_arr = np.zeros(data_arr.shape[0])
28
29             # for elements [0, 1, 2, ... ema-1] fill 1
               element, 2 element, ... averages
30             for i in range(ema):
31                 ema_arr[i] = np.mean(data_arr[0:(i+1)])
32
33             ema_arr[ema] = np.mean(data_arr[0:ema])
34             for i in range(ema, data_arr.shape[0]):
```

```
35              ema_arr[i] = ((ema-1)*ema_arr[i-1] + data_
                arr[i])/float(ema)
36              if np.isnan(ema_arr[i]):
37                  ema_arr[i] = np.mean(data_arr[i-ema:i])
38
39          return ema_arr
40
41
42     class MktModel(object):
43         DAYS_IN_WEEK = 5
44
45         def __init__(self, dr):
46             mkt_file = os.path.join(dr, "SP500.csv")
47             self.df = pd.read_csv(mkt_file)
48             self.df.loc[:, "Date"] = pd.to_datetime(self.
                df.loc[:, "Date"])
49             self.confVal = 0.95
50             self.df = self.calculateVars()
51
52         def calculateVars(self):
53             df = self.df
54             px = df.loc[:, "Adj Close"].values
55             rows = df.shape[0]
56             ret = np.log(np.divide(px[self.DAYS_IN_
                WEEK-1:-1], px[0:rows-self.DAYS_IN_WEEK]))
57             df.loc[:, "MktReturn"] = 0.0
58             df.loc[self.DAYS_IN_WEEK:, "MktReturn"] = ret
59             # volatility of returns
60             df.loc[:, "MktVolatility"] = 0.0
61             mvolat = np.zeros(df.shape[0], dtype=np.float64)
62             mvol = np.zeros(df.shape[0], dtype=np.float64)
```

```
63          avgVol = np.mean(df.Volume.
            values[0:int(rows*0.7)])
64          for i in range(self.DAYS_IN_WEEK, df.shape[0]):
65              mvolat[i] = np.std(df.loc[i - self.DAYS_IN_
                WEEK:i - 1, "MktReturn"])
66              mvol[i] = np.sum(df.loc[i - self.DAYS_IN_
                WEEK:i - 1, "Volume"].values) / avgVol
67          df.loc[:, "MktVolatility"] = mvolat
68          df.loc[:, "MktVolume"] = mvol
69          return df
70
71      def buildModel(self, fname=None):
72          df = self.df
73          ret = df.loc[self.DAYS_IN_WEEK:,
            "MktReturn"].values
74          # build a AR model
75          pacf, confint = statsmodels.tsa.stattools.
            pacf(ret, alpha=0.05)
76          # plot pacf, confint
77          fig, ax = plt.subplots()
78          #fig.suptitle("AR Model Identification")
79          y_err = np.subtract(confint, np.reshape(np.
            repeat(pacf, 2), confint.shape))
80          xpos = np.arange(len(pacf))
81          ax.bar(xpos[1:], pacf[1:], yerr=y_err[1:, 1],
            alpha=0.5, ecolor="black", capsize=2)
82          ax.set_title("AR Model Identification")
83          ax.set(ylabel='PACF')
84          ax.set(xlabel="Order")
85          #ax.set_xticks(xpos)
86          ax.yaxis.grid(True)
```

```
87              #axs[1].set(ylabel="Conf Int")
88              plt.tight_layout()
89              plt.show()
90              plt.close(fig)
91              self.df = df
92              mod = self.buildOrder5Model(df)
93              return mod, pacf, confint
94
95      def buildOrder5Model(self, df):
96              vals = df.loc[self.DAYS_IN_WEEK:,
                "MktReturn"].values
97              laggedvals = [vals[0:-i*self.DAYS_IN_WEEK] for i
                in range(1, 6)]
98
99              x_data = sm.add_constant(np.
                vstack([laggedvals[0][4*self.DAYS_IN_WEEK:],
100              laggedvals[1][3*self.DAYS_IN_WEEK:],
101                  laggedvals[2][2*self.DAYS_IN_WEEK:],
102                      laggedvals[3][1*self.DAYS_IN_WEEK:],
103                      laggedvals[4],
104      df.MktVolatility.values[5*self.DAYS_IN_WEEK:-self.DAYS_
        IN_WEEK],
105                  df.MktVolume.values[5*self.DAYS_IN_WEEK:-
                self.DAYS_IN_WEEK]]).T)
106              lm_model = lm.OLS(vals[5*self.DAYS_IN_
                WEEK:], x_data)
107              result = lm_model.fit()
108              # check p values for significance
109              print("R^2 = %f" % result.rsquared_adj)
110              for pval in result.pvalues:
111                  if pval > (1 - self.confVal):
```

```
112                 print("Values are not significant at 95%
                    significance level")
113         self.df = df
114         return result
115
116
117     class SectorModel(object):
118         def __init__(self, dir_name, sector, mkt_df):
119             sct_file = os.path.join(dir_name, "%s.
                csv"%sector)
120             self.df = self.readData(sct_file, mkt_df)
121             self.mktDf = mkt_df
122             self.confVal = 0.95
123
124         def readData(self, sct_file, mkt_df):
125             df = pd.read_csv(sct_file)
126             df.loc[:, "Date"] = pd.to_datetime(df.loc[:,
                "Date"])
127             vals = df.loc[:, "Adj Close"].values
128             ret = np.log(np.divide(vals[MktModel.DAYS_IN_
                WEEK-1:-1], vals[0:-MktModel.DAYS_IN_WEEK]))
129             df.loc[:, "Return"] = 0.0
130             df.loc[MktModel.DAYS_IN_WEEK:, "Return"] = ret
131
132             config = ColumnConfig()
133             ema_10 = TransformedRiskMeasure('PxEMA10')
134             df.loc[:, ema_10.name] = ema_10.
                calculateEMA(df[config.CLOSE_PRICE].
                values, ema=10)
135             ema_20 = TransformedRiskMeasure('PxEMA20')
```

```
136        df.loc[:, ema_20.name] = ema_20.
           calculateEMA(df[config.CLOSE_PRICE].
           values, ema=20)
137        df.loc[:, "ShortMLong"] = np.where(df.loc[:,
           ema_10.name].values > df.loc[:, ema_20.name].
           values, 1, 0)
138
139        df.loc[:, "ActReturn"] = 0.0
140        df.loc[0:df.shape[0]-MktModel.DAYS_IN_WEEK-1,
           "ActReturn"] = ret
141
142        mkt_df.rename(columns={"Adj Close": "MktPx"},
           inplace=True)
143        df = pd.merge(df, mkt_df[["Date", "MktReturn",
           "MktPx", "MktVolume", "MktVolatility"]],
           on=["Date"], how="inner")
144        return df
145
146    def buildModel(self):
147        df = self.df
148        ret = df.loc[:, "Return"].values
149        mktret = df.loc[:, "MktReturn"].values
150        laggedret = ret[MktModel.DAYS_IN_WEEK:-MktModel.
           DAYS_IN_WEEK]
151        df.loc[:, "LaggedReturn"] = 0.0
152        df.loc[2*MktModel.DAYS_IN_WEEK:, "LaggedReturn"]
           = laggedret
153
154        x_data = sm.add_constant(np.vstack([mktret
           [3*MktModel.DAYS_IN_WEEK:],
155                laggedret[0:-MktModel.DAYS_IN_WEEK]]).T)
```

```
156         lm_model = lm.OLS(ret[3*MktModel.DAYS_IN_
            WEEK:], x_data)
157         result = lm_model.fit()
158         # check p values for significance
159         print("R^2 = %f" % result.rsquared_adj)
160         for pval in result.pvalues:
161             if pval > (1 - self.confVal):
162                 print("Values are not significant at 95%
                    significance level")
163
164         return result
165
166
167 class LSTMModel(object):
168     def __init__(self, df, training_data_perc=0.70,
        validation_data_perc=0.05, symbol='',
169                 return_sequences=True):
170         self.symbol = symbol
171         self.returnSequences = return_sequences
172         self.nTimeSteps = 4
173         rows = df.shape[0]
174         trg_begin = 0
175         trg_end = int(training_data_perc * rows)
176         validation_begin = trg_end + 1
177         validation_end = int((training_data_perc +
            validation_data_perc) * rows)
178         self.df = df
179         x_train, y_train = self.getTrainingData(df.
            loc[trg_begin:trg_end, :].reset_
            index(drop=True))
```

```
180          x_valid, y_valid = self.getValidationData(df.
             loc[validation_begin:validation_end, :].reset_
             index(drop=True))
181          self.lstm = self.buildLSTMModel(x_train,
             y_train, x_valid, y_valid)
182
183      def getTrainingData(self, df):
184          data_arr = df.loc[:, ["MktVolatility",
             "MktReturn", "MktVolume", "Return"]].values
185          actret_arr = df.loc[:, "ActReturn"].values
186          input_arr = np.zeros((data_arr.
             shape[0]-5*MktModel.DAYS_IN_WEEK, self.
             nTimeSteps, 4), dtype=np.float64)
187          if self.returnSequences:
188              output_arr = np.zeros((input_arr.shape[0],
                 self.nTimeSteps))
189          else:
190              output_arr = np.zeros(input_arr.shape[0])
191          debug_df = pd.DataFrame(data={"Date": df.Date})
192          lcols = ["L%d"%i for i in range(self.
             nTimeSteps-1, -1, -1)]
193          cols = list(itertools.product(lcols,
             ["MktVolatility", "MktReturn", "MktVolume",
             "Return"]))
194          cols = [c[0]+c[1] for c in cols]
195          cols2 = ["L%dActReturn"%i for i in range(self.
             nTimeSteps-1, -1, -1)]
196          for cl1 in cols + cols2:
197              debug_df.loc[:, cl1] = 0.0
198          offset = 4*MktModel.DAYS_IN_WEEK
```

```
199          for i in range(offset, data_arr.shape[0]-
             MktModel.DAYS_IN_WEEK):
200              for j in range(self.nTimeSteps):
201                  input_arr[i-offset, j, :] = data_arr[i-
                     (self.nTimeSteps-1-j)*MktModel.DAYS_IN_
                     WEEK, :]
202              debug_df.loc[i, cols] = input_arr[i-offset,
                 :, :].flatten()
203              if self.returnSequences:
204                  for j in range(self.nTimeSteps):
205                      output_arr[i-offset, j] =
                         actret_arr[i-(self.nTimeSteps-1-
                         j)*MktModel.DAYS_IN_WEEK]
206                  debug_df.loc[i, cols2] = output_arr[i-
                     offset, :]
207              else:
208                  output_arr[i - offset] = actret_arr[i]
209          df_final = pd.merge(df, debug_df, on=["Date"],
             how="left")
210          return input_arr, output_arr
211
212      def getValidationData(self, df):
213          data_arr = df.loc[:, ["MktVolatility",
             "MktReturn", "MktVolume", "Return"]].values
214          actret_arr = df.loc[:, "ActReturn"].values
215          if data_arr.shape[0] <= 5*MktModel.DAYS_IN_WEEK:
216              return None, None
217          input_arr = np.zeros((data_arr.shape[0] -
             5*MktModel.DAYS_IN_WEEK, self.nTimeSteps, 4),
             dtype=np.float64)
218          if self.returnSequences:
```

```
219             output_arr = np.zeros((input_arr.shape[0],
                self.nTimeSteps))
220         else:
221             output_arr = np.zeros(input_arr.shape[0])
222         offset = 4*MktModel.DAYS_IN_WEEK
223         for i in range(offset, data_arr.shape[0] -
            MktModel.DAYS_IN_WEEK):
224             for j in range(self.nTimeSteps):
225                 input_arr[i - offset, j, :] = data_
                    arr[i - (self.nTimeSteps-1-j)*MktModel.
                    DAYS_IN_WEEK, :]
226             if self.returnSequences:
227                 for j in range(self.nTimeSteps):
228                     output_arr[i - offset, j] =
                        actret_arr[i - (self.nTimeSteps-1-
                        j)*MktModel.DAYS_IN_WEEK]
229             else:
230                 output_arr[i - offset] = actret_arr[i]
231         return input_arr, output_arr
232
233     def buildLSTMModel(self, x_train, y_train, x_valid,
        y_valid):
234         model = models.Sequential()
235         lyr = layers.LSTM(8, return_sequences=self.
            returnSequences)
236         #lyr = tf.keras.layers.SimpleRNN(8, return_
            sequences=self.returnSequences)
237         model.add(lyr)
238         model.add(layers.Dense(4))
239         model.add(layers.Dense(1))
240         input_shape = (None, self.nTimeSteps, 4)
```

```
241         model.build(input_shape)
242         model.summary()
243         model.compile(optimizer="adam", loss=tf.keras.
            losses.MeanSquaredError(), metrics=["mse"])
244         if x_valid is not None:
245             model.fit(x_train, y_train, validation_
                data=(x_valid, y_valid), epochs=5)
246         else:
247             model.fit(x_train, y_train, epochs=5)
248         return model
249
250     def predict(self, df, begin):
251         lcols = ["L%d" % i for i in range(self.
            nTimeSteps - 1, -1, -1)]
252         cols = list(itertools.product(lcols,
            ["MktVolatility", "MktReturn", "MktVolume",
            "Return"]))
253         cols2 = ["L%dActReturn"%i for i in range(self.
            nTimeSteps-1, -1, -1)]
254         if self.returnSequences:
255             cols3 = ["L%dPrReturn"%i for i in
                range(self.nTimeSteps-1, -1, -1)]
256         else:
257             cols3 = ["L0PrReturn"]
258         cols = [c[0] + c[1] for c in cols]
259         data_arr = df.loc[begin:, ["MktVolatility",
            "MktReturn", "MktVolume", "Return"]].values
260         actret_arr = df.loc[begin:, "ActReturn"].values
261         results_df = pd.DataFrame(data={"Date":
            df.loc[begin:, "Date"]})
262         for cl1 in cols+cols2+cols3:
```

```
263                         results_df.loc[:, cl1] = 0.0
264                 input_arr = np.zeros((1, self.nTimeSteps, 4),
                    dtype=np.float64)
265                 output_arr = np.zeros((1, self.nTimeSteps),
                    dtype=np.float64)
266                 for i in range(begin + 3*MktModel.DAYS_IN_WEEK,
                    df.shape[0]-MktModel.DAYS_IN_WEEK):
267                     for j in range(self.nTimeSteps):
268                         input_arr[0, j, :] = data_arr[i -
                            begin - (self.nTimeSteps-1-j)*MktModel.
                            DAYS_IN_WEEK, :]
269                     results_df.loc[i, cols] = input_arr[0, :,
                        :].flatten()
270                     for j in range(self.nTimeSteps):
271                         output_arr[0, j] = actret_arr[i -
                            begin - (self.nTimeSteps-1-j)*MktModel.
                            DAYS_IN_WEEK]
272                     results_df.loc[i, cols2] = output_arr[0, :]
273                     out1 = self.lstm.predict(input_arr)
274                     results_df.loc[i, cols3] = out1.flatten()
275             results_df = pd.merge(results_df, df,
                on=["Date"], how="left")
276             return results_df
277
278         @staticmethod
279         def plot(df, begin, secname, fname=None):
280             fig, ax = plt.subplots(nrows=1, ncols=3)
281             fig.set_size_inches((30, 7), forward=True)
282             column = "Adj Close"
283             ylabel = "Price"
284             end = df.shape[0] - MktModel.DAYS_IN_WEEK
```

```
285         dates = df.Date[begin:end+1].values
286         majorLocator = YearLocator()  # every year
287         minorLocator = MonthLocator()  # every month
288         formatter = DateFormatter('%Y')
289
290         ax[0].plot(dates, df.loc[begin:end,
            column].values)
291         ax[0].set_ylabel(ylabel)
292         ax[0].xaxis.set_major_locator(majorLocator)
293         ax[0].xaxis.set_major_formatter(formatter)
294         ax[0].xaxis.set_minor_locator(minorLocator)
295         ax[0].format_xdata = DateFormatter('%Y-%m')
296         ax[0].set_xlabel("Date")
297         ax[0].grid(True)
298
299         columns = ["LOPrReturn", "ActReturn"]
300         ylabels = ["Pr. Return", "Ac. Return"]
301         for i in range(1, 3):
302             ax[i].bar(dates, df.loc[begin:end,
                columns[i-1]].values, alpha=0.5,
                ecolor="black")
303             ax[i].set_ylabel(ylabels[i-1])
304             ax[i].xaxis.set_major_locator(majorLocator)
305             ax[i].xaxis.set_major_formatter(formatter)
306             ax[i].xaxis.set_minor_locator(minorLocator)
307             ax[i].format_xdata = DateFormatter('%Y-%m')
308             ax[i].set_xlabel("Date")
309             ax[i].grid(True)
310
311         plt.title(secname)
312         #plt.tight_layout()
```

```
313              plt.show()
314              plt.close(fig)
315
316
317    class RegressionModelPredictor(object):
318        def __init__(self, dir_name, sector):
319            mkt_coeff = os.path.join(dir_name, "mkt.csv")
320            self.mktDf = pd.read_csv(mkt_coeff)
321            sector_coeff = os.path.join(dir_name,
                   "coeff.csv")
322            self.sectorDf = pd.read_csv(sector_coeff)
323            self.sectorDf = self.sectorDf.loc[self.sectorDf.
                   Sector.eq(sector), :].reset_index(drop=True)
324
325        def predict(self, df, begin):
326            mkt_lags = len(self.mktDf.columns) - 3
327            mkt_x = np.zeros(mkt_lags + 2, dtype=np.float64)
328            mktret = df.MktReturn.values
329            mktvolat = df.MktVolatility.values
330            mktvolume = df.MktVolume.values
331            secret = df.Return.values
332            df.loc[:, "RegPrReturn"] = 0.0
333            cols = ["const"] + ["L%d" % i for i in range(1,
                   mkt_lags + 1)] + ["MktVolatility", "MktVolume"]
334            mkt_coeff = self.mktDf.loc[0, cols].values
335            sec_x = np.zeros(2, dtype=np.float64)
336            sec_coeff = self.sectorDf.loc[0, ["const",
                   "MktReturn", "LaggedReturn"]].values
337            for i in range(begin + 3 * MktModel.DAYS_IN_
                   WEEK, df.shape[0] - MktModel.DAYS_IN_WEEK):
338                for j in range(mkt_lags):
```

```
339                    mkt_x[j] = mktret[i - j * MktModel.DAYS_
                       IN_WEEK]
340                mkt_x[mkt_lags] = mktvolat[i]
341                mkt_x[mkt_lags+1] = mktvolume[i]
342                pred_ret = mkt_coeff[0] + np.dot(mkt_
                   coeff[1:], mkt_x)
343                sec_x[0] = pred_ret
344                sec_x[1] = secret[i]
345                pred_sec_ret = sec_coeff[0] + np.dot(sec_
                   coeff[1:], sec_x)
346                df.loc[i, "RegPrReturn"] = pred_sec_ret
347            df, rms_reg, rms_lstm = self.sqDiff(df, begin)
348            return df, rms_reg, rms_lstm
349
350        @staticmethod
351        def plot(df, begin, fname=None, sec=''):
352            fig, ax = plt.subplots(nrows=3, ncols=1)
353            end = df.shape[0] - MktModel.DAYS_IN_WEEK
354            dates = df.Date[begin:end + 1].values
355            majorLocator = YearLocator()  # every year
356            minorLocator = MonthLocator()  # every month
357            formatter = DateFormatter('%Y')
358
359            cols = ["RegPrReturn", "SqRegDiff",
                   "SqLSTMDiff"]
360            ylabels = ["Reg. Pr. Return", "Sq. Diff.",
                   "Sq. Diff."]
361            for i in range(3):
362                ax[i].bar(dates, df.loc[begin:end, cols[i]].
                       values, alpha=0.5, ecolor="black")
363                ax[i].set_ylabel(ylabels[i])
```

```
364                   ax[i].xaxis.set_major_locator(majorLocator)
365                   ax[i].xaxis.set_major_formatter(formatter)
366                   ax[i].xaxis.set_minor_locator(minorLocator)
367                   ax[i].format_xdata = DateFormatter('%Y-%m')
368                   ax[i].set_xlabel("Date")
369                   ax[i].grid(True)
370                   #ax[i].title.set_text(sec)
371
372              fig.suptitle(sec)
373              plt.show()
374              plt.close(fig)
375
376         def sqDiff(self, df, begin):
377              df.loc[:, "SqRegDiff"] = 0.0
378              df.loc[:, "SqLSTMDiff"] = 0.0
379              nr = df.shape[0]
380              diff = np.subtract(df.loc[begin:,
                   "RegPrReturn"].values, df.loc[begin:,
                   "ActReturn"].values)
381              df.loc[begin:, "SqRegDiff"] =
                   np.multiply(diff, diff)
382              diff = np.subtract(df.loc[begin:, "LOPrReturn"].
                   values, df.loc[begin:, "ActReturn"].values)
383              df.loc[begin:, "SqLSTMDiff"] =
                   np.multiply(diff, diff)
384              avg_rmsreg = np.sqrt(np.sum(df.loc[begin:nr -
                   MktModel.DAYS_IN_WEEK, "SqRegDiff"].values) /
                   (nr - begin - MktModel.DAYS_IN_WEEK))
385              avg_rmslstm = np.sqrt(np.sum(df.loc[begin:nr -
                   MktModel.DAYS_IN_WEEK, "SqLSTMDiff"].values) /
                   (nr - begin - MktModel.DAYS_IN_WEEK))
```

```
386                 return df, avg_rmsreg, avg_rmslstm
387
388         def trade(self, df, begin):
389             sgs = ["RegSignal", "LSTMSignal"]
390             cols = ["RegPrReturn", "LOPrReturn"]
391             for signal, col in zip(sgs, cols):
392                 df.loc[:, signal] = 0
393                 skip = 0
394                 last_pos = 0
395                 for i in range(begin, df.shape[0] -
                    MktModel.DAYS_IN_WEEK):
396                     if skip > i:
397                         continue
398                     if last_pos == 0:
399                         if df.loc[i, col] > 0:
400                             last_pos = 1
401                             df.loc[i, signal] = 1
402                     elif last_pos == 1:
403                         if df.loc[i, col] < 0:
404                             last_pos = 0
405                             df.loc[i, signal] = -1
406                     else:
407                         raise ValueError("Invalid value of
                        last_pos: %d"%last_pos)
408                 if last_pos == 1:
409                     df.loc[df.shape[0] - MktModel.DAYS_IN_
                    WEEK, signal] = -1
410             return df
411
412
413     def regression(input_dir, output_dir):
```

```
414        dir_name = input_dir
415        model = MktModel(dir_name)
416        pacf_file = os.path.join(output_dir, "mkt_pacf.png")
417        vals = model.buildModel(pacf_file)
418        mkt_params = vals[0].params
419        sfilename = os.path.join(output_dir, "summary.txt")
420        sfile = open(sfilename, "w")
421
422        sfile.write(vals[0].summary().as_text())
423        df1 = pd.DataFrame(data={"const": [mkt_params[0]],
           "L1": [mkt_params[1]], "L2": [mkt_params[2]],
424                                       "L3": [mkt_params[3]],
                                          "L4": [mkt_params[4]],
                                          "L5": [mkt_params[5]],
425                                       "MktVolatility": [mkt_
                                          params[6]], "MktVolume":
                                          [mkt_params[7]]})
426        coeff_file = os.path.join(output_dir, "mkt.csv")
427        df1.to_csv(coeff_file, index=False)
428        """
429        Communication services: XLC
430        Consumer Discretionary: XLY
431        Consumer Staples: XLP
432        Energy: XLE
433        Financials: XLF
434        Healthcare: XLV
435        Industrials: XLI
436        Information Technology: XLK
437        Materials: XLB
438        Real Estate: XLRE
439        Utilities: XLU
440        """
```

```
441        sectors = ["XLC", "XLY", "XLP", "XLE", "XLF", "XLV",
           "XLI", "XLK", "XLB", "XLRE", "XLU"]
442        results = pd.DataFrame(data={"Sector": sectors})
443        results.loc[:, "const"] = 0.0
444        results.loc[:, "MktReturn"] = 0.0
445        results.loc[:, "LaggedReturn"] = 0.0
446        for sec in sectors:
447            smodel = SectorModel(dir_name, sec, model.df)
448            res = smodel.buildModel()
449            sfile.write("\n" + sec + "\n")
450            sfile.write(res.summary().as_text())
451            params = res.params
452            row = results.Sector.eq(sec)
453            results.loc[row, "const"] = params[0]
454            results.loc[row, "MktReturn"] = params[1]
455            results.loc[row, "LaggedReturn"] = params[2]
456        coeff_file = os.path.join(output_dir, "coeff.csv")
457        results.to_csv(coeff_file, index=False)
458        sfile.close()
459
460
461    def runLSTM(input_dir, output_dir):
462        dir_name = input_dir
463        model = MktModel(dir_name)
464        sectors = ["XLC", "XLY", "XLP", "XLE", "XLF", "XLV",
           "XLI", "XLK", "XLB", "XLRE", "XLU"]
465        return_seq = False
466        for sec in sectors:
467            smodel = SectorModel(dir_name, sec, model.df)
468            avg_vol = np.mean(smodel.df.Volume.
           values[0:int(0.75 * smodel.df.shape[0])])
```

```
469            lstm = LSTMModel(smodel.df, symbol=sec, return_
               sequences=return_seq)
470            begin = int(0.75 * smodel.df.shape[0])
471            result_df = lstm.predict(smodel.df, begin)
472            result_df.to_csv(os.path.join(output_dir, "%s_
               lstmpredict.csv"%sec))
473            plot_file = os.path.join(output_dir, "%s_plots.
               png" % sec)
474            lstm.plot(result_df, 0, sec, plot_file)
475
476    def plotLSTMResults(input_dir, output_dir):
477        dir_name = input_dir
478        sectors = ["XLC", "XLY", "XLP", "XLE", "XLF", "XLV",
               "XLI", "XLK", "XLB", "XLRE", "XLU"]
479        rmsDf = pd.DataFrame(data={"Sector": sectors,
               "RMSReg": [0]*len(sectors), "RMSLSTM":
               [0]*len(sectors)})
480        for sec in sectors:
481            sec_file = os.path.join(dir_name, "%s.csv"%sec)
482            df = pd.read_csv(sec_file)
483            avg_vol = np.mean(df.Volume.values[0:int(0.75 *
               df.shape[0])])
484            fl = os.path.join(output_dir, "%s_lstmpredict.
               csv"%sec)
485            df = pd.read_csv(fl)
486            plot_file = os.path.join(output_dir, "%s_plots.
               png"%sec)
487            LSTMModel.plot(df, 0, sec, plot_file)
488            rpred = RegressionModelPredictor(output_
               dir, sec)
489            df, rms1, rms2 = rpred.predict(df, 0)
```

```
490              print("Sector: %s, RMSReg %f, RMSLSTM %f" %
                 (sec, rms1, rms2))
491              rmsDf.loc[rmsDf.Sector.eq(sec), "RMSReg"] = rms1
492              rmsDf.loc[rmsDf.Sector.eq(sec),
                 "RMSLSTM"] = rms2
493              df = rpred.trade(df, 0)
494              reg_plot = os.path.join(output_dir, "reg_%s_
                 plots.png"%sec)
495              rpred.plot(df, 0, reg_plot, sec)
496
497          rmsDf.to_csv(os.path.join(output_dir, "rmserr.csv"))
498          print(rmsDf.to_latex(index=False))
499
500      if __name__ == "__main__":
501          input_dir = r"C:\prog\cygwin\home\samit_000\value_
                 momentum_new\value_momentum\data\sectors"
502          output_dir = r"C:\prog\cygwin\home\samit_000\value_
                 momentum_new\value_momentum\output\sector"
503          regression(input_dir, output_dir)
504          runLSTM(input_dir, output_dir)
505          plotLSTMResults(input_dir, output_dir)
```

Once trained, the two models are used to predict 1-week returns from 2015 to 2020 for each sector. To compare their performance, let us look at plots of standard deviation of weekly returns predicted by the model from the actual returns observed. The plots show that the two models produce similar results. Standard deviation of predicted returns from the actual returns for the two models has been shown in Table 4-3. As can be seen, the values for the two models are close, with the regression model showing marginally better prediction. For XLE, the LSTM model gives a better prediction (in minimum root-mean-square sense). This demonstrates the effectiveness of the LSTM model in identifying correlations in data. The actual numbers obtained for the LSTM model may vary slightly across runs due to random weight initialization.

***Table 4-3.*** *Comparison of Root Mean Square Error Between Predicted and Actual Returns*

| Sector | RMS Error (Reg.) | RMS Error (LSTM) |
|--------|------------------|-------------------|
| XLC    | 0.043592         | 0.099679          |
| XLY    | 0.024379         | 0.026290          |
| XLP    | 0.018359         | 0.024590          |
| XLE    | 0.037844         | 0.037125          |
| XLF    | 0.031782         | 0.040148          |
| XLV    | 0.022252         | 0.022622          |
| XLI    | 0.026149         | 0.026444          |
| XLK    | 0.025065         | 0.031712          |
| XLB    | 0.026835         | 0.028032          |
| XLRE   | 0.043028         | 0.045240          |
| XLU    | 0.024322         | 0.025932          |

# CHAPTER 5

# Reinforcement Learning Theory

This chapter lays out basic reinforcement learning theory. It introduces the notation used in reinforcement learning literature and provides detailed explanation and proofs of underlying concepts. It provides the foundation for reinforcement learning algorithms introduced in the next chapter.

Richard Bellman pioneered the development of reinforcement learning in the 1950s (Dreyfus, 2002) with the formulation of the Bellman equation governing the optimal state-action selection in a Markov decision problem (MDP). Most researchers applied dynamic programming for solving the Bellman equation – an approach that suffered from the curse of dimensionality and the fact that it required a model of system dynamics. Due to intractability of this approach and unavailability of a model governing system dynamics for most problems, approximation methods began to emerge. In 1989, in a seminal paper titled "Sequential Decision Problems and Neural Networks," Andrew G. Barto, Richard S. Sutton, and Chris Watkins advocated the use of TD (temporal difference) learning methods as a means of combining learning and optimal selection in the Bellman equation. With the development of sophisticated networks over the following two decades, neural networks began to be used as policy and value functions in reinforcement learning. After 2010, several groundbreaking applications of reinforcement learning emerged where

S. Ahlawat, *Reinforcement Learning for Finance*,
https://doi.org/10.1007/978-1-4842-8835-1_5

a reinforcement learning agent was able to outperform human actors. DDQN, A3C, DDPG, and dueling DDPG – to name just a few – are examples of algorithms that have achieved great success in their fields of application.

# 5.1 Basics

Reinforcement learning is a category of learning algorithms within artificial intelligence that learn from a history of rewards earned by taking an action prescribed by a policy with the objective of maximizing the sum of expected discounted future rewards. Unlike supervised learning, it does not require a set of labels (classification) or true values (regression) for learning. It learns from prior experience of rewards with the objective of maximizing the sum of expected future discounted rewards. Furthermore, many algorithms within reinforcement learning do not require a model of the environment. Intuitively, reinforcement learning is akin to a child learning complex actions like how to be successful at school from rewards and punishments for simple actions like doing homework on time. In a supervised learning framework, one would have to teach a child on how to be successful by showing them examples of other children who did things a certain way and achieved success. As one can readily observe, the number of examples (training data) required for such a training effort to be effective would be impractically large. Consider all the desirable qualities (independent variables) that have a bearing on academic success such as attending classes, being punctual, higher education, and so on and their permutations. No parent or educator would keep such extensive records of students. This is an illustration of the problem of the curse of dimensionality. However, even if the problem of training data paucity is surmountable, the child would quickly lose interest in attempting to learn from examples because they may question the relevance of those examples. Different circumstances of certain students in the training data items may render those data points inapposite. This illustrates the problem of unavailability of model dynamics: the child is unsure which factors are the primary drivers of academic success in their circumstances.

A reinforcement learning problem consists of an environment, an agent, and a policy. It is formulated as a Markov decision problem (MDP). All dynamics within a Markov decision problem are governed by the current state and action. Historical states and actions have no bearing on system dynamics. An environment is an abstraction for the process that monitors the state of the agent, accepts actions, distributes a reward, and transitions to a next state. An agent represents the learner that seeks to learn a policy. A policy is a generic rule that prescribes which actions to take in a certain state. A policy can be stochastic, in which case there is a probability distribution for each action in a given state. The objective of reinforcement learning is to make the agent learn a policy in order to maximize the sum of expected future discounted rewards. MDP is a tuple $(\mathcal{S}, \mathcal{A}, \mathcal{P}, \mathcal{R}, \gamma)$. Let us use the following notation to describe the MDP:

1. Let $\mathcal{S}$ denote the set of states and $\mathcal{A}$ denote the set of actions. These sets can be continuous or discreet.

2. Let $s_t$ denote the state of the environment at time $t$, with $s_t \in \mathcal{S}$.

3. $a_t$ denotes the action of the agent at time $t$ with $a_t \in \mathcal{A}$.

4. $\mathcal{R}(s_{t+1}, s_t, a_t)$ denotes the reward process. In general, it could be a function of the next state, current state, and action. For environments with a deterministic state transition function, that is, where $s_t$ and $a_t$ determine $s_{t+1}$, $R$ is a function of $s_t$ and $a_t$ only.

5. $r_t$ denotes the reward at time $t$.

6. $\gamma$ is the discount factor for weighing future rewards. For applications where future rewards are less valuable than immediate rewards, this factor is less than 1. This factor needs to be less than 1 for applications with infinite time horizon and non-zero rewards. For problems with finite time horizon, $\gamma$ can be 1. In general, $\gamma \in [0, 1]$.

7. Let $\mathcal{P}(s_{t+1}|s_t, a_t) \in \mathbb{R}$ denote the state transition function. For a stochastic state transition function, this is a real number with $\sum_{s_{t+1}} \mathcal{P}(s_{t+1}|s_t, a_t) = 1$. For a deterministic state transition function, $\mathcal{P}(s_{t+1}|s_t, a_t) = \delta_{s_{t+1}, \tilde{s}}$, where $\delta_{i,j}$ is the Kronecker delta symbol with the property shown in equation 5.1. $\tilde{s}$ is the deterministic state that follows the occurrence of action $a_t$ in state $s_t$:

$$\delta_{ij} = \begin{cases} 1, & \text{if } i = j, \\ 0, & \text{if } i \neq j. \end{cases} \tag{5.1}$$

8. Let $p_0(s)$ denote the probability of the agent being in state $s$ at initial time $t_0$. We have $\sum_{s \in \mathcal{S}} p_0(s) = 1$.

9. $\pi(a_t|s_t) \in \mathbb{R}$ denotes the policy prescribing the action to take in state $s_t$. This could be stochastic. A deterministic policy prescribes one action for a given state. Hence, deterministic policies are represented as $\pi(s_t)$.

10. The total discounted reward following a policy is given by equation 5.2:

$$J^\pi = \sum_{t_0}^{\infty} \gamma^t r(s_t, a_t)$$
$$= \sum_{s \in S} d^\pi(s) \sum_{a \in A} \pi(a|s) r(s,a) \qquad (5.2)$$

$d^\pi(s|s_0)$ is the discounted stationary probability distribution of states under policy $\pi(a_t|s_t)$ and the state transition function $p(s_{t+1}|s_t, a_t)$. In other words, $d^\pi(s|s_0)$ is the discounted probability of being in a state $s$ at any time, starting from state $s_0$. It can be written as the sum of combined probabilities of visiting a state $s$ at any time step, as shown in equation 5.3:

$$d^\pi(s|s_0) = \sum_{t=0}^{\infty} \gamma^t P(s_t = s|s_0, \pi)$$
$$= p_0(s) + \gamma \sum_{a_0 \in A} \sum_{s_0 \in S} p_0(s_0) \pi(a_0|s_0) P(s_1|s_0, a_0) +$$
$$\gamma^2 \sum_{a_0 \in A} \sum_{a_1 \in A} \sum_{s_0 \in S} \sum_{s_1 \in S} p_0(s_0) \pi(a_0|s_0) P(s_1|s_0, a_0)$$
$$\pi(a_1|s_1) P(s_2|s_1, a_1) + \cdots \qquad (5.3)$$

11. The state-action value function (Q function) for a policy $\pi$ is the reward obtained by taking an action in a state and following the policy in subsequent steps, as shown in equation 5.4:

$$Q^\pi(s_t, a_t) = \sum_{s_{t+1} \in S} P(S_{t+1}|S_t, a_t) r(S_{t+1}, S_t, a_t) +$$
$$\gamma \sum_{s_{t+1} \in S} \sum_{a_{t+1} \in A} P(S_{t+1}|S_t, a_t) \pi(a_{t+1}|S_{t+1}) Q^\pi(S_{t+1}, a_{t+1})$$
$$= E_{s_{t+1}} \left[ r(s_{t+1}, s_t, a_t) + \gamma \sum_{a_{t+1} \in A} \pi(a_{t+1}|S_{t+1}) Q^\pi(S_{t+1}, a_{t+1}) \right]$$
$$= E_{s_{t+1}} \left[ r(s_{t+1}, s_t, a_t) + \gamma E_{a_{t+1}} \left[ \pi(a_{t+1}|S_{t+1}) Q^\pi(S_{t+1}, a_{t+1}) \right] \right] \qquad (5.4)$$

For environments with a deterministic state transition function, state $s_t$ and action $a_t$ determine the next state $s_{t+1}$. For such environments, the state-action value function $Q^\pi(s_t, a_t)$ can be written as shown in equation 5.5:

$$Q^\pi\left(s_t, a_t\right) = r\left(s_t, a_t\right) + \gamma E_{a_{t+1}}\left[\pi\left(a_{t+1}|s_{t+1}\right)Q^\pi\left(s_{t+1}, a_{t+1}\right)\right] \tag{5.5}$$

It is often convenient to sample from the model in order to get an expected value instead of taking the actual expectation over the state transition function. In case a model is unavailable, we assume that the sampled episode gives a sample from the underlying but unknown Markov model. In this setting, the Q function can be written as shown in equation 5.6:

$$Q^\pi\left(s_t, a_t\right) = r\left(s_t, a_t\right) + \gamma E_{a_{t+1} \sim \pi(\cdot|s_{t+1})}\left[Q^\pi\left(s_{t+1}, a_{t+1}\right)\right] \tag{5.6}$$

12. The state value function of a policy is the average reward earned in a state by following a policy, as shown in equation 5.7:

$$\begin{aligned} V^\pi\left(s_t\right) &= \sum_{a_t \in A}\pi\left(a_t|s_t\right)Q^\pi\left(s_t, a_t\right) \\ &= E_{a_t \sim \pi(\cdot|s_t)}\left[Q^\pi\left(s_t, a_t\right)\right] \end{aligned} \tag{5.7}$$

Using the state value function, equation for the state-action value function $Q^\pi(s_t, a_t)$ can be simplified to equation 5.8:

$$Q^\pi\left(s_t, a_t\right) = E_{s_{t+1}}\left[r\left(s_{t+1}, s_t', a_t\right) + \gamma V^\pi\left(s_{t+1}\right)\right] \tag{5.8}$$

Substituting equation 5.8 in equation 5.7, we obtain equation 5.9:

$$
\begin{aligned}
V^{\pi}\left(s_{t}\right) &= \sum_{a_{t} \in A} \pi\left(a_{t} \mid s_{t}\right) E_{s_{t+1}}\left[r\left(s_{t+1}, s_{t}, a_{t}\right)+\gamma V^{\pi}\left(s_{t+1}\right)\right] \\
&= E_{a_{t}, s_{t+1}}\left[r\left(s_{t+1}, s_{t}, a_{t}\right)+\gamma V^{\pi}\left(s_{t+1}\right)\right] \\
&= E_{a_{t}, s_{t+1}, a_{t+1}, s_{t+2}, \cdots}\left[r\left(s_{t+1}, s_{t}, a_{t}\right)+\gamma r\left(s_{t+2}, s_{t+1}, a_{t+1}\right)+\right. \\
&\quad \left. \gamma^{2} r\left(s_{t+3}, s_{t+2}, a_{t+2}\right)+\cdots\right]
\end{aligned}
\tag{5.9}
$$

Equation 5.9 illustrates why $V$ is the average sum of rewards obtained by following a policy $\pi$. For a deterministic policy, an action is fully prescribed by the policy as a function of state, and equation 5.9 can be simplified to equation 5.10:

$$
\begin{aligned}
V_{\text{det.Policy}}^{\pi}\left(s_{t}\right) &= E_{s_{t+1}, s_{t+2}, \cdots}\left[r\left(s_{t+1}, s_{t}, \pi\left(s_{t}\right)\right)\right. \\
&\quad \left.+\gamma r\left(s_{t+2}, s_{t+1}, \pi\left(s_{t+1}\right)\right)+\gamma^{2} r\left(s_{t+3}, s_{t+2}, \pi\left(s_{t+2}\right)\right)+\cdots\right]
\end{aligned}
\tag{5.10}
$$

13. An advantage function represents the improvement in the state-action value over the value function in a state by following an action, as shown in equation 5.11:

$$
A^{\pi}\left(s_{t}, a_{t}\right)=E_{s_{t+1}}\left[Q^{\pi}\left(s_{t}, a_{t}\right)-V^{\pi}\left(s_{t}\right)\right]
\tag{5.11}
$$

14. When using stochastic sampling from distribution prescribed by the state transition function, an advantage function can be written as shown in equation 5.12:

$$
A^{\pi}\left(s_{t}, a_{t}\right)=r\left(s_{t+1}, s_{t}, a_{t}\right)+\gamma V^{\pi}\left(s_{t+1}\right)-V^{\pi}\left(s_{t}\right)
\tag{5.12}
$$

## 5.2 Methods for Estimating the Markov Decision Problem

Markov decision problem (MDP) estimation can be done using supervised learning or reinforcement learning. Supervised learning methods learn the state transition function $P(s_{t+1}|s_t, a_t)$ and the reward function $R(s_{t+1}, s_t, a_t)$ using methods such as the hidden Markov model (HMM). These methods represent the state transition function and reward function using parametric functions and then learn the model parameters. The second method of estimating MDP is reinforcement learning. Since this book focuses on reinforcement learning, we will only look at the latter category of estimation methods.

Reinforcement learning methods for estimating MDP can be grouped into value function approximation methods, policy approximation methods, and actor-critic methods. We look at each of these methods in the following sections. A pictorial depiction of categorization of reinforcement learning algorithms can be seen in Figure 5-1.

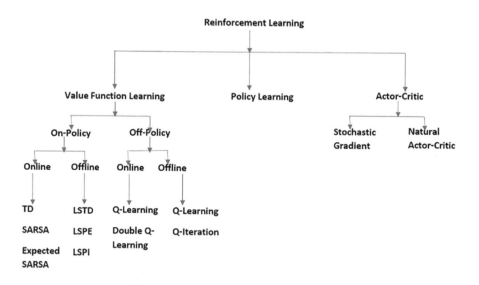

Categorization of Reinforcement Learning Algorithms

*Figure 5-1.*  *Reinforcement Learning Algorithms*

# 5.3 Value Estimation Methods

These methods learn the state-action value function from the reward experience. The goal of reinforcement learning is to learn an optimal policy for the agent. An optimal policy in these methods is inferred from the state-action value function. The Bellman equation for both the state value function and state-action value function must be satisfied for a consistent value function, as shown in equation 5.13:

$$V^{\pi}(s) = E_{a \sim \pi, s'}\left[r(s', s, a) + \gamma V^{\pi}(s')\right]$$
$$Q^{\pi}(s, a) = E_{s'}\left[r(s', s, a) + \gamma \sum_{a' \in A} \pi(a'|s')Q^{\pi}(s', a')\right]$$

$$(5.13)$$

For an optimal policy $\pi^*$, the Bellman equation in equation 5.14 must be satisfied. $Q * (s_t, a_t)$ denotes the state-action value function corresponding to the optimal policy. With this function at hand, a deterministic optimal policy can be obtained using equation 5.15. Using the optimal policy, an optimal value function can be written as shown in equation 5.16:

$$Q^*\left(s_t,a_t\right)=E_{s_{t+1}}\left[r\left(s_{t+1},s_t,a_t\right)+\gamma\max_{a'}Q^*\left(s_{t+1},a'\right)\right] \tag{5.14}$$

$$\pi^*\left(s_t\right)=\operatorname*{argmax}_{a'\in\mathcal{A}}Q^*\left(s_t,a'\right) \tag{5.15}$$

$$V^*\left(s_t\right)=E_{s_{t+1}}\left[r(s_{t+1},s_t,\pi^*\left(s_t\right))+\gamma V^*\left(s_{t+1}\right)\right] \tag{5.16}$$

There are three general methods for solving the Bellman equation: dynamic programming, Monte Carlo methods, and TD learning. While dynamic programming requires a model of the environment, Monte Carlo and TD learning are model-free methods and do not require a model of the environment. Let us look at each of these methods.

## 5.3.1 Dynamic Programming

Dynamic programming solves a problem by partitioning it into smaller ones, recursively solving the smaller ones and putting the solutions together to solve the original problem. The Bellman equation shown in equation 5.14 is amenable to solution by dynamic programming if we have the model of the environment. Specifically, we require the state transition function $P(s_{t+1}|s_t,a_t)$ and the reward function $R(s_{t+1},s_t,a_t)$. The state transition function is required to calculate the expectation, and the reward function gives the reward. However, for most problems, a model of the environment is not available.

# Finding the Optimal Path in a Maze

Let us look at an example of solving the Bellman equation using dynamic programming. We have a maze, as shown in Figure 5-2. The objective is to enable the agent to find the shortest path from the entry to the exit square. The squares shown in black represent walls and cannot be traversed. At each step, the agent can move up, down, left, or right subject to the condition that the landing square is not a wall or outside the maze.

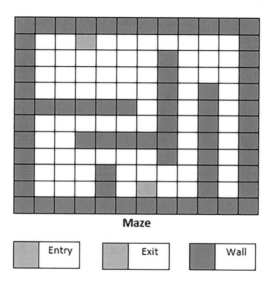

**Maze**

| | Entry | | | Exit | | | Wall |

***Figure 5-2.*** *Maze*

In this problem, state is the current position of the agent. The state transition function is deterministic and is completely determined by the existing state (position) and action of the agent. Action space is discreet with four choices: left, right, up, or down for the next move, subject to the constraints.

Let us formulate a reward function. Upon reaching the exit square, the agent gets a reward of 1. In order to ensure that the selected policy picks the shortest path from entry to exit, moves to all squares other than the exit have a reward of –1. Reward for moving into walls can be assigned a value of $-\infty$ to ensure that the agent never steps on them. The objective of the problem is to find a policy for the agent that maximizes the total reward. The discount factor has a value of 1, since future rewards are as valuable as present rewards.

To ensure that the agent does not cross the boundary, let us create two additional rows and columns of squares bounding the maze and consider them as walls. A table for $Q^*(s_t, a_t)$ is created with dimensions (12 × 12, 4). Size of state space, $\mathcal{S}$, is 12 × 12 because there are 10 + 2 rows and 10 + 2 columns in the maze, and there are four actions for each square, subject to constraints. $Q*(\text{wall}, a) = -\infty$ for all actions $a$. Let us also set $Q*(\text{exit}, a) = 1$. All other $Q^*$ values for admissible states are initialized to a large negative number, signifying that they have not been calculated yet. The initial values are shown in equation 5.17. Now the problem is fully formulated and is amenable to solving using dynamic programming:

$$Q^*(\text{wall}, a) = -\infty$$
$$Q^*(\text{exit}, a) = 1$$

(5.17)

The Bellman equation is applied to calculate the value of the state-action value function $Q^*$ for each state and action combination. If the calculation encounters a state-action pair whose $Q^*$ value has not been calculated yet, a recursive call is made. It is important to detect cycles in this process. This is done using a set, *seenSet* in the code shown. Once we have obtained $Q^*(s, a)$, we can find the optimal policy, that is, the shortest path from any square to the exit square using equation 5.15. The full code is shown in Listing 5-1, and the selected path is shown in Figure 5-3. The time complexity of the algorithm is $\Theta(12 \times 12 \times 4 \times 4)$, and the space complexity is $\Theta(12 \times 12 \times 4)$.

***Listing 5-1.*** Solving the Maze Problem Using Dynamic
Programming

```
 1   import numpy as np
 2   from enum import Enum, unique
 3   import logging
 4
 5   logging.basicConfig(level = logging.INFO)
 6   logger = logging.getLogger(__name__)
 7
 8   @unique
 9   class Actions(Enum):
10       LEFT = 0
11       RIGHT = 1
12       UP = 2
13       DOWN = 3
14
15
16   class MazeSolver(object):
17       NEG_INFTY = float(-1E10)
18       NOT_SET = float(-1E8)
19
20       def __init__(self, entry, exit):
21           self.gamma = 1.0
22           self.mazeSize = (12, 12)
23           self.nActions = len(Actions)
24           self.entry = entry
25           self.exit = exit
26           self.walls = {(5,1), (5,2), (5,3), (5,4), (5,5),
27                         (7,3), (7,4), (7,5), (7,6),
28                         (2,7), (3,7), (4,7), (5,7), (6,7),
                          (7,7), (8,7),
```

```
29                        (4,9), (5,9), (6,9), (7,9), (8,9),
                          (9,9), (10,9),
30                        (9,4), (10,4)}
31          self.actionMap = {Actions.LEFT : (-1, 0),
32                            Actions.RIGHT : (1, 0),
33                            Actions.UP : (0, -1),
34                            Actions.DOWN : (0, 1)}
35          # add bounding walls
36          for i in range(self.mazeSize[0]):
37              self.walls.add((i, 0))
38              self.walls.add((i, self.mazeSize[1]-1))
39
40          for j in range(self.mazeSize[1]):
41              self.walls.add((0, j))
42              self.walls.add((self.mazeSize[0]-1, j))
43          if self.entry in self.walls:
44              raise ValueError("Entry square is
                  inadmissible")
45          if self.exit in self.walls:
46              raise ValueError("Exit square is
                  inadmissible")
47          self.QStar = np.ndarray((self.mazeSize[0], self.
                mazeSize[1], self.nActions), dtype=np.float)
48          self.initQStar(self.QStar)
49
50      def transitionFunc(self, state0, action):
51          increments = self.actionMap[action]
52          return state0[0] + increments[0], state0[1] +
                increments[1]
53
```

```
54      def rewardFunc(self, state0, action):
55          state1 = self.transitionFunc(state0, action)
56          if state1 in self.walls:
57              return MazeSolver.NEG_INFTY
58          elif state1 == self.exit:
59              return 1
60          return -1
61
62      def initQStar(self, Q):
63          for i in range(self.mazeSize[0]):
64              for j in range(self.mazeSize[1]):
65                  square = (i,j)
66                  if square in self.walls:
67                      for action in Actions:
68                          Q[i, j, action.value] =
                            MazeSolver.NEG_INFTY
69                  else:
70                      for action in Actions:
71                          Q[i, j, action.value] =
                            MazeSolver.NOT_SET
72
73          for action in Actions:
74              Q[self.exit[0], self.exit[1], action.
                value] = 0
75
76      def dpBellman(self, state, action, seenSet=None):
77          # returns Q(state, action)
78          if self.QStar[state[0], state[1], action.value]
                != MazeSolver.NOT_SET:
79              return self.QStar[state[0], state[1],
                action.value]
80
```

```
81          if seenSet is None:
82              seenSet = {(state[0], state[1],
                action.value)}
83          elif (state[0], state[1], action.value) in
            seenSet:
84              # cycle detected, backtrack, so other paths
                can be explored
85              return MazeSolver.NEG_INFTY
86
87          reward = self.rewardFunc(state, action)
88          if reward == MazeSolver.NEG_INFTY:
89              self.QStar[state[0], state[1], action.value]
                = MazeSolver.NEG_INFTY
90              return MazeSolver.NEG_INFTY
91
92          seenSet.add((state[0], state[1], action.value))
93          nextstate = self.transitionFunc(state, action)
94          maxval = MazeSolver.NEG_INFTY
95          for aprime in Actions:
96              val = self.dpBellman(nextstate, aprime,
                seenSet)
97              if val > maxval:
98                  maxval = val
99          if maxval == MazeSolver.NEG_INFTY:
100             self.QStar[state[0], state[1], action.value]
                = MazeSolver.NEG_INFTY
101             return MazeSolver.NEG_INFTY
102
103         self.QStar[state[0], state[1], action.value] =
            reward + self.gamma * maxval
```

```
104            return self.QStar[state[0], state[1],
               action.value]
105
106        def optPolicy(self):
107            for action in Actions:
108                self.dpBellman(self.entry, action)
109
110            optpath = [self.entry]
111            sq = self.entry
112            while sq != self.exit:
113                maxval = MazeSolver.NEG_INFTY
114                bestaction = None
115                for action in Actions:
116                    if maxval < self.QStar[sq[0], sq[1],
                       action.value]:
117                        bestaction = action
118                        maxval = self.QStar[sq[0], sq[1],
                           action.value]
119
120                if bestaction is None:
121                    return optpath
122                sq = self.transitionFunc(sq, bestaction)
123                optpath.append(sq)
124
125            return optpath
126
127
128    if __name__ == "__main__":
129        entry = (1, 3)
130        exit = (10, 6)
131        maze_solver = MazeSolver(entry, exit)
```

```
132        path = maze_solver.optPolicy()
133        if path[-1] != exit:
134            logger.info("No path exists")
135
136        logger.info("->".join([str(p) for p in path]))
```

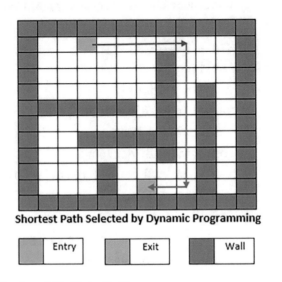

**Shortest Path Selected by Dynamic Programming**

Entry    Exit    Wall

***Figure 5-3.*** *Optimal Path to Exit*

The output path produced by the code is shown in Listing 5-2.

***Listing 5-2.*** Maze Path to Exit

```
1    (1, 3)->(1, 4)->(1, 5)->(1, 6)->(1, 7)->(1, 8)->(2, 8)->
(3, 8)->(4, 8)->(5, 8)->(6, 8)->(7, 8)->(8, 8)->(9, 8)->
(10, 8)->(10, 7)->(10, 6)
```

# European Call Option Valuation

A European call option is a financial instrument that gives the holder
the right but not the obligation to buy a specific asset at strike price *K* at
maturity *T* of the contract. The option can only be exercised at maturity.

Let us denote the underlying asset's price at time $t$ by $S_t$, risk-free rate by $r_f$, and volatility of the underlying asset by $\sigma$. If the volatility of asset $\sigma$ and risk-free rate $r_f$ are assumed to be constant, the asset is assumed to not pay the dividend, and the asset price is assumed to follow log-normal dynamics as shown in equation 5.18. Price of a European call option $V_t$ is given by the Black-Scholes formula shown in equation 5.19. The price does not depend on the asset's rate of return $\mu$, because one can create a risk-free portfolio comprised of the call option and $-\dfrac{\partial V_t}{\partial S_t}$ units of the underlying asset.

$$dS_t = \mu S_t dt + \sigma S_t dW_t$$
$$dW_t = \epsilon \sqrt{dt} \tag{5.18}$$
$$\epsilon \sim \text{Standard Normal Distribution}$$

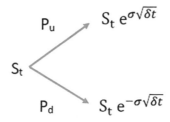

**Figure 5-4.**  *Stock Price Recombining Grid*

$$V_t = N(d_1)S_t - N(d_2)Ke^{-r_f(T-t)}$$
$$N(x) = \int_{-\infty}^{x} \frac{1}{2\pi} e^{-\frac{y^2}{2}} dy$$
$$d_1 = \frac{\ln\dfrac{S_t}{K} + \left( r_f + \dfrac{\sigma^2}{2} \right)(T-t)}{\sigma\sqrt{T-t}} \tag{5.19}$$
$$d_2 = d_1 - \sigma\sqrt{T-t}$$

Let us use dynamic programming to compute the price of a European call option. Let us first discretize the state space. State space consists of stock price on the Y axis and time steps on the X axis. At each time step, stock price can move up to $Se^{\sigma\sqrt{(\Delta t)}}$ with probability $P_u$, or it can fall down to $Se^{-\sigma\sqrt{(\Delta t)}}$ with probability $P_d$ as shown in Figure 5-4.

In order to keep the grid recombining, we must ensure that going up at time $t$ followed by going down at time $t + 1$ ends up in the same node as going down at $t$ followed by going up at $t + 1$. A recombining grid is more tractable computationally because it has a linearly growing number of states, whereas a non-recombining grid has an exponentially growing number of states. For option pricing, the asset is assumed to have a drift $\mu$ equal to the risk-free rate $r_f$. The underlying asset's price equation satisfies the log-normal equation, whose solution is shown in equation 5.20:

$$dS_t = r_f S_t dt + \sigma S_t dW_t$$

$$S_T = S_t e^{\left(r_f - \frac{\sigma^2}{2}\right)(T-t)} e^{\sigma\sqrt{T-t\epsilon}}$$

(5.20)

The time step must be chosen small enough so that the expected price of an asset at time $t + \Delta t$ is between $Se^{-\sigma\sqrt{\Delta t}}$ and $Se^{\sigma\sqrt{\Delta t}}$, as shown in equation 5.21:

$$E[S_{t+\Delta t}] = Se^{\left(r_f + \frac{\sigma^2}{2}\right)\Delta t}$$

$$Se^{-\sigma\sqrt{\Delta t}} \le Se^{\left(r_f + \frac{\sigma^2}{2}\right)\Delta t} \le Se^{\sigma\sqrt{\Delta t}}$$

$$\Rightarrow -\sigma \le \left(r_f + \frac{\sigma^2}{2}\right)\sqrt{\Delta t} \le \sigma$$

(5.21)

$$\Rightarrow \Delta t \le \left(\frac{\sigma}{r_f + \frac{\sigma^2}{2}}\right)^2$$

Probabilities of moving up $P_u$ and down $P_d$ are chosen to match the expected stock price at time $t + \Delta t$ as shown in equation 5.22:

$$P_u + P_d = 1$$

$$P_u S_t e^{\sigma \sqrt{\Delta t}} + P_d S_t e^{-\sigma \sqrt{\Delta t}} = E[S_{t+\Delta t}]$$

$$= S_t e^{\left(r_f + \frac{\sigma^2}{2}\right)\Delta t}$$

$$P_u = \frac{e^{\left(r_f + \frac{\sigma^2}{2}\right)\Delta t} - e^{-\sigma \sqrt{\Delta t}}}{e^{\sigma \sqrt{\Delta t}} - e^{-\sigma \sqrt{\Delta t}}}$$

$$P_d = 1 - P_u$$

(5.22)

Having discretized the stock price–time domain, let us define the value function as the price of the option at node $(S_t, t)$ to be $V(S_t, t)$. This can be written as a function of the price at nodes at time step $t + \Delta t$ as shown in equation 5.23:

$$V(S_t, t) = e^{-r_f \Delta t}(P_u V(S_u, t + \Delta t) + P_d V(S_d, t + \Delta t)) \tag{5.23}$$

At maturity, option price is 0 if the stock price is below strike price $K$ and $S_T - K$ if it is above the strike price, as shown in equation 5.24:

$$V(S_T, T) = \begin{cases} S_T - K & \text{if } S_T \geq K \\ 0 & \text{otherwise} \end{cases} \tag{5.24}$$

Let us consider a European call option on a publicly traded stock with time to maturity $T$ of 2 months, risk-free rate $r_f$ of 0.5% per annum, volatility of stock $\sigma$ to be 20% per annum, moneyness or ratio of strike price to stock price $\frac{K}{S_0}$ to be 1.1, and current stock price $S_0$ to be $20. Using the Black-Scholes formula, the price of this option is 0.1048.

Using dynamic programming as described previously, this option's price can be calculated using code shown in Listing 5-3. As seen from the output, the calculated option value of 0.103 is close to the Black-Scholes price.

***Listing 5-3.*** Calculating a European Call Option's Price Using
Dynamic Programming

```
1    import numpy as np
2    import logging
3    from scipy.stats import norm
4
5    logging.basicConfig(level=logging.DEBUG)
6    logger = logging.getLogger("root")
7
8
9    class EuropeanOption(object):
10       def __init__(self, s0, strike, maturity, rf,
         volatility, minsteps=20):
11           """
12           Initialize
13           :param s0: Initial price of underlying asset
14           :param strike: Strike price
15           :param maturity: Maturity in years
16           :param rf: Risk free rate (per annum)
17           :param volatility: expressed per annum
18           :param minsteps: Minimum number of time steps
19           """
20           self.s0 = s0
21           self.strike = strike
22           self.maturity = maturity
23           self.rf = rf
24           self.vol = volatility
25           self.minSteps = minsteps
26
```

```
27          self.deltaT = min(self.calculateDeltaT(),
            maturity/minsteps)
28          self.df = np.exp(-rf * self.deltaT)
29          self.sqrtTime = np.sqrt(self.deltaT)
30          expected = np.exp((rf +
            volatility*volatility/2.0)*self.deltaT)
31          self.up = np.exp(volatility * self.sqrtTime)
32          self.down = np.exp(-volatility * self.sqrtTime)
33          self.pUp = (expected - self.down)/(self.up -
            self.down)
34          self.pDown = 1.0 - self.pUp
35          self.ntime = int(np.ceil(maturity / self.deltaT))
36          self.grid = np.zeros((2*self.ntime, self.ntime),
            dtype=np.float32)
37
38      def evaluate(self):
39          # values at time T
40          grid = self.grid
41          val = self.s0 * np.exp(-volatility * self.sqrtTime
            * self.ntime)
42          for i in range(2*self.ntime):
43              grid[i, -1] = max(val - self.strike, 0)
44              val *= self.up
45
46          for j in range(self.ntime-1, 0, -1):
47              for i in range(self.ntime-j, self.ntime+j, 1):
48                  grid[i, j-1] = self.df * (self.pUp *
                    grid[i+1, j] + self.pDown * grid[i-1, j])
49
50          return grid[self.ntime, 0]
51
```

```
52        def calculateDeltaT(self):
53            val = self.vol / (self.rf + self.vol*self.vol/2.0)
54            return val*val
55
56        def blackScholes(self):
57            d1 = (np.log(self.s0/self.strike) +
58                  (self.rf + self.vol*self.vol/2.0)*self.
                  maturity)/(self.vol * np.sqrt(self.
                  maturity))
59            d2 = d1 - self.vol * np.sqrt(self.maturity)
60            return self.s0 * norm.cdf(d1) - self.strike *
                  np.exp(-self.rf * self.maturity) * norm.cdf(d2)
61
62
63    if __name__ == "__main__":
64        price = 20.0
65        strike = 22.0
66        maturity = 2.0/12.0
67        volatility = 0.2
68        rf = 0.005
69        eoption = EuropeanOption(price, strike, maturity, rf,
              volatility, minsteps=25)
70        bsPrice = eoption.blackScholes()
71        simPrice = eoption.evaluate()
72        logger.info("Black Scholes price: %f, simulated price:
              %f", bsPrice, simPrice)
```

Output from the code can be seen in Listing 5-4.

***Listing 5-4.*** Computed Option Price

```
1   Black Scholes price: 0.104751, simulated price: 0.103036
```

# Valuation of a European Barrier Option

Barrier options are a class of exotic options whose payoff depends on the price of the underlying asset hitting a barrier. There are two classes of barrier options, each of which is further subdivided into two types, as described in the following:

1. A **knock-in** barrier option is worthless unless the asset price reaches or crosses a barrier value. This option is subdivided into the following two types:

    a. An **up-and-in** barrier option acquires value only if the underlying asset price crosses the barrier from below, that is, the price becomes equal or exceeds the barrier prior to the option's maturity.

    b. A **down-and-in** barrier option has a non-zero value only if the underlying asset price reaches or falls below a barrier prior to the option's maturity.

2. A **knock-out** barrier option becomes worthless if the underlying asset price reaches or crosses a barrier value. Like its knock-in counterpart, this option also has two subtypes:

    a. An **up-and-out** option becomes worthless if the underlying asset's price reaches or exceeds a barrier.

    b. A **down-and-out** option becomes worthless if the underlying asset's price reaches or falls below a barrier.

A European barrier option can only be exercised at maturity and is similar in other respects to its American counterpart. Barrier options were discussed in an earlier section. A European option is less valuable than its corresponding American option. Because there is no early-exercise feature, we do not need to use the state-action value function.

As before, represent the state value function using a two-dimensional (price, time) grid. Let us consider a European knock-in barrier call option on a publicly traded stock with barrier $B$ of \$23, time to maturity $T$ of 2 months, risk-free rate $r_f$ of 0.5% per annum, volatility of stock $\sigma$ to be 20% per annum, moneyness or ratio of strike price to stock price $\dfrac{K}{S_0}$ to be 1.1, and current stock price $S_0$ to be \$20. This implies the strike price $K$ is \$22.

Let $P_h(S_t, t)$ denote the probability of stock price hitting the barrier from below and reaching price $S_t$ at time $t$. This can be written as shown in equation 5.25. $P_u$ and $P_d$ are the probabilities of stock price moving up or down from the current price obtained from equation 5.22. Equation 5.25 can be understood as follows: If the underlying asset's price $S_t$ is greater than or equal to the barrier price $B$, $P_h(S_t, t) = 1$. If not, $P_h(S_t, t)$ is equal to the probability of hitting the barrier en route to the previous upper node and moving down or the previous lower node and moving up to reach the current node at time $t$.

$$P_h\left(S_t,t\right)=\begin{cases}1 \text{ if } S_t \geq B \\ P_d P_h\left(\dfrac{S_t}{e^{-\sigma\sqrt{\Delta t}}},t-\Delta t\right)+P_u P_h\left(\dfrac{S_t}{e^{\sigma\sqrt{\Delta t}}},t-\Delta t\right)\text{otherwise}\end{cases} \tag{5.25}$$

Value of the option is the probability weighted discounted price at nodes in the next time step as shown in equation 5.26. There is no early-exercise feature in a European option. Because the option has value only if it has hit the barrier from below, equation 5.26 has a multiplier $P_h(S_t, t)$ to account for this condition. Similarly, the barrier hitting probability must be backed out of the value function at $(S_u, t + \Delta t)$ and $(S_d, t + \Delta t)$ nodes because those nodes are being visited from the $(S_t, t)$ node.

$$V(S_t,t) = e^{-r_f \Delta T} P_h(S_t,t) \left( P_u \frac{V(S_u,t+\Delta t)}{P_h(S_u,t+\Delta t)} + P_d \frac{V(S_d,t+\Delta t)}{P_h(S_d,t+\Delta t)} \right) \qquad (5.26)$$

At maturity, option price is given by equation 5.27. $P_h(S_T, T)$ is the probability of the price having hit the barrier from below as calculated using equation 5.25.

$$V(S_T,T) = P_h(S_T,T) \max(S_T - K, 0) \qquad (5.27)$$

The dynamic programming code for valuing this option is shown in Listing 5-5. The option price is around 0.0041 – less than the price of the plain vanilla European call option computed in the previous section. The reduction in price is due to the additional barrier constraint that may cause the option to expire worthless.

***Listing 5-5.*** Calculating a European Barrier Up-and-In Call Option's Price Using Dynamic Programming

```
1    import numpy as np
2    import logging
3    import matplotlib.pyplot as plt
4    from mpl_toolkits.mplot3d import Axes3D
5
6    logging.basicConfig(level=logging.DEBUG)
7    logger = logging.getLogger("root")
8
9
10   class EuropeanKnockInCallOption(object):
11       def __init__(self, so, strike, maturity, rf,
             volatility, barrier, minsteps=20):
```

```
12              """
13              Initialize
14              :param s0: Initial price of underlying asset
15              :param strike: Strike price
16              :param maturity: Maturity in years
17              :param rf: Risk free rate (per annum)
18              :param volatility: expressed per annum
19              :param barrier: Barrier for this knock-in option
20              :param minsteps: Minimum number of time steps
21              """
22              self.s0 = s0
23              self.strike = strike
24              self.barrier = barrier
25              self.maturity = maturity
26              self.rf = rf
27              self.vol = volatility
28              self.minSteps = minsteps
29
30              self.deltaT = min(self.calculateDeltaT(),
                maturity/minsteps)
31              self.df = np.exp(-rf * self.deltaT)
32              self.sqrtTime = np.sqrt(self.deltaT)
33              expected = np.exp((rf +
                volatility*volatility/2.0)*self.deltaT)
34              self.up = np.exp(volatility * self.sqrtTime)
35              self.down = np.exp(-volatility * self.sqrtTime)
36              self.pUp = (expected - self.down)/(self.up -
                self.down)
37              self.pDown = 1.0 - self.pUp
38              self.ntime = int(np.ceil(maturity / self.deltaT))
```

```
39          self.grid = np.zeros((2*self.ntime, self.ntime),
            dtype=np.float32)
40          self.price = None
41          self.hitProb = self.calcBarrierHitProb()
42
43      def calcBarrierHitProb(self):
44          # calculate probability for t=0
45          hitprob = np.zeros((2*self.ntime, self.ntime),
            dtype=np.float32)
46          price = np.full(self.ntime*2, self.up, dtype=np.
            float32)
47          price[0] = self.s0 * (self.down ** self.ntime)
48          price = np.cumprod(price)
49          self.price = price
50
51          hitprob[:, -1] = np.where(price >= self.barrier,
            1.0, 0.0)
52
53          # for t = 1, 2, ... ntime-1
54          for j in range(self.ntime-2, -1, -1):
55              for i in range(self.ntime-j, self.ntime+j+1):
56                  if price[i] >= self.barrier:
57                      hitprob[i, j] = 1.0
58                  else:
59                      hitprob[i, j] = self.pUp *
                        hitprob[i+1, j+1] + self.pDown *
                        hitprob[i-1, j+1]
60          return hitprob
61
62      def evaluate(self):
63          # values at time T
```

```
64              grid = self.grid
65              val = self.s0 * np.exp(-volatility * self.
                sqrtTime * self.ntime)
66              for i in range(2*self.ntime):
67                  grid[i, -1] = self.hitProb[i, -1] * max(val -
                    self.strike, 0)
68                  val *= self.up
69
70              for j in range(self.ntime-1, 0, -1):
71                  for i in range(self.ntime-j, self.
                    ntime+j, 1):
72                      val1 = 0
73                      if self.hitProb[i+1, j] > 0:
74                          val1 = grid[i+1, j]/self.
                            hitProb[i+1, j]
75                      val2 = 0
76                      if self.hitProb[i-1, j] > 0:
77                          val2 = grid[i-1, j]/self.
                            hitProb[i-1, j]
78                      grid[i, j-1] = self.df * self.hitProb[i,
                        j-1] * (self.pUp * val1 + self.
                        pDown * val2)
79
80          return grid[self.ntime, 0]
81
82      def calculateDeltaT(self):
83          val = self.vol / (self.rf + self.vol*self.
            vol/2.0)
84          return val*val
85
86      def plotPrice(self):
```

```
87              price = self.price
88              time = np.full(self.ntime, self.deltaT, dtype=np.
                float32)
89              time[0] = 0
90              time = np.cumsum(time)
91              x, y = np.meshgrid(price, time)
92              fig = plt.figure()
93              axs = fig.add_subplot(111, projection='3d')
94              axs.plot_surface(x.T, y.T, self.grid)
95              axs.set_xlabel('Stock Price')
96              axs.set_ylabel('Time (Yrs)')
97              axs.set_zlabel('Option Price')
98              plt.show()
99
100             fig, axs = plt.subplots(1, 1, constrained_
                layout=True)
101             cs = axs.contourf(x.T, y.T, self.grid)
102             fig.colorbar(cs, ax=axs, shrink=0.85)
103             axs.set_title("European Barrier Knock-In Call
                Option")
104             axs.set_ylabel("Time to Maturity (yrs)")
105             axs.set_xlabel("Initial Stock Price")
106             axs.locator_params(nbins=5)
107             axs.clabel(cs, fmt="%1.1f", inline=True,
                fontsize=10, colors='w')
108             plt.show()
109
110
111     if __name__ == "__main__":
112         price = 20.0
113         strike = 22.0
```

```
114      maturity = 2.0/12.0
115      barrier = 23.0
116      volatility = 0.2
117      rf = 0.005
118      eoption = EuropeanKnockInCallOption(price, strike,
         maturity, rf, volatility, barrier, minsteps=25)
119      simPrice = eoption.evaluate()
120      logger.info("simulated price: %f", simPrice)
121      eoption.plotPrice()
```

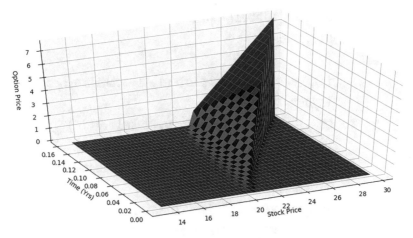

***Figure 5-5.*** *Option Price Surface for a European Barrier Up-and-In Call Option*

The option price surface is shown in Figure 5-5 against stock price and time to maturity (in years). Contour plot of the option price is shown in Figure 5-6. As seen in the plots, the price of the knock-in option is close to 0 near option maturity and below the barrier at $23.

# 5.3.2 Generalized Policy Iteration

If a model of the environment is available, generalized policy iteration can be used to find an optimal policy. Generalized policy iteration involves value function estimation using an initial policy followed by greedy improvement of the policy. The process is repeated using the improved policy until it converges to an optimal policy. Before delving into policy iteration, let us look at the policy improvement theorem, which provides the foundation for establishing convergence of policy iteration.

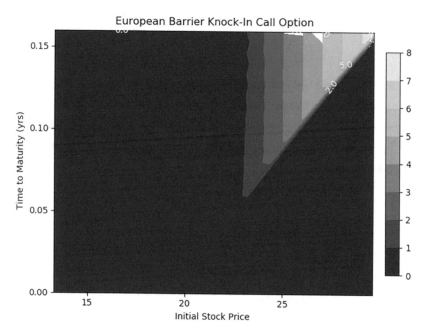

***Figure 5-6.*** *Option Price Contour Plot for a European Barrier Up-and-In Call Option*

## Policy Improvement Theorem

According to the policy improvement theorem, any two deterministic policies $\pi$ and $\pi^*$ that satisfy the condition in equation 5.28 for all states $s \in \mathcal{S}$, $\pi^*$ are a better policy than $\pi$:

$$Q^\pi\left(s,a=\pi^*(s)\right)\geq V^\pi(s) \text{ for all } s \in \mathcal{S} \tag{5.28}$$

If a policy $\pi^*$ is better than $\pi$, it necessarily implies that the value function obtained using $\pi^*$ has a higher value than that obtained using $\pi$ for all states $s \in \mathcal{S}$.

$$V^{\pi^*}(s)\geq V^\pi(s) \text{ for all } s \in \mathcal{S} \tag{5.29}$$

In order to prove the policy improvement theorem, let us expand the action value function in condition 5.28 using equation 5.8. The policy being deterministic, action is prescribed by the policy at each step. It should be noted that the state transition function can be stochastic and need not be deterministic.

$$
\begin{aligned}
V^\pi(s) &\leq Q^\pi\left(s,a=\pi^*(s)\right) \\
&= E_{s'}\left[r(s',s,\pi^*(s))+\gamma V^\pi(s')\right] \\
&\leq E_{s'}\left[r(s',s,\pi^*(s))+\gamma Q^\pi(s',\pi^*(s'))\right] \\
&= E_{s',s''}\left[r(s',s,\pi^*(s))+\gamma r(s'',s',\pi^*(s'))+\gamma^2 V^\pi(s'')\right] \\
&\cdots \\
&= E_{s',s'',s''',\dots}\left[r(s',s,\pi^*(s))+\gamma r(s'',s',\pi^*(s'))+\gamma r(s''',s'',\pi^*(s''))+\dots\right] \\
&= V^{\pi^*}(s)
\end{aligned}
\tag{5.30}
$$

The last equation in equation 5.30 follows as a result of the expression for the state value function for a deterministic policy, as shown in equation 5.10. Finally, $V^\pi(s)\leq V^{\pi^*}(s)$ for all states $s \in \mathcal{S}$ implies that $\pi^*$ is a better policy (or at least, as good a policy for the case of equality) than $\pi$.

# Policy Evaluation

Policy evaluation is the evaluation of the state value function or state-action value function using a specified policy. If the model of the environment is known, one can explicitly use equation 5.9 and equation 5.4 to calculate the state value function and action value function, respectively. An iterative algorithm for calculating these functions is illustrated below. The underlying principle of the algorithms is that as the value function converges to the true value function, iterative corrections will become zero.

Let us look at an iterative calculation of the state value function. Equation 5.31 is used to iteratively update the state value function for each state. The algorithm is shown in pseudo-code 1.

$$V^{\pi}(s_{t}) = E_{a_{t}, s_{t+1}}\left[r(s_{t+1}, s_{t}, a_{t}) + \gamma V^{\pi}(s_{t+1})\right] \tag{5.31}$$

---

**Algorithm 1**    Iterative Policy Evaluation Algorithm for Computing the Value Function

**Require:**    Policy $\pi(s, a)$ and model of environment $p(s' | s, a)$ and $r(s', s, a)$

1: Initialize $V(s)$ to 0, for all $s \in \mathcal{S}$.

2: **repeat**

3:    $\Delta \leftarrow 0$

4:    **for** each $s \in \mathcal{S}$ **do**

5:        $v \leftarrow V(s)$

6:        $V(s) \leftarrow \sum_{a} \pi(s, a) \sum_{s' \in \mathcal{S}} p(s' | s, a)\left[r(s', s, a) + \gamma V(s')\right]$

7:      $\Delta \leftarrow \max\left(\Delta, |v - V(s)|\right)$

8:  **end for**

9: **until** $\Delta < \in$

---

The state-action value function can be evaluated for a specific policy using a similar approach. Equation 5.32 is used to update the action value function, $Q(s, a)$. The full algorithm is presented in pseudo-code 2.

$$Q^{\pi}(s, a) = E_{s'}\left[ r(s', s, a) + \gamma \sum_{a' \in \mathcal{A}} \pi(a'|s')Q^{\pi}(s', a') \right] \qquad (5.32)$$

---

**Algorithm 2**  Iterative Policy Evaluation Algorithm for Computing the State-Action Value Function

**Require:**  Policy $\pi(s, a)$ and model of environment $p(s'|s, a)$ and $r(s', s, a)$

1: Initialize $Q(s, a) = 0$, for all $s \in \mathcal{S}$ and $a \in \mathcal{A}$.

2: **repeat** $\Delta \leftarrow 0 \ q \leftarrow Q(s, a)$

3:  **for** each $s \in \mathcal{S}$ **do**

4:    **for** each $a \in \mathcal{A}$ **do**

5:      $q \leftarrow Q(s, a)$

6:      $Q(s, a) \leftarrow \sum_{s' \in \mathcal{S}} p(s'|s, a)\left[ r(s', s, a) + \gamma \sum_{a' \in \mathcal{A}} \pi(a'|s')Q(s', a') \right]$

7:      $\Delta \leftarrow \max\left(\Delta, |q - Q(s, a)|\right)$

8:    **end for**

9:  **end for**

10: **until** $\Delta < \in$

---

# Policy Improvement

Policy improvement involves using a state-action value function to improve the policy that generated it. Policy improvement is achieved by using greedy action selection at each state giving the maximum value of the action function, as shown in equation 5.33:

$$\pi'(s) = \operatorname*{argmax}_{a' \in \mathcal{A}} Q^{\pi}(s, a') \forall s \in \mathcal{S} \tag{5.33}$$

$\pi'(s)$ being a deterministic policy, equation 5.34 necessarily follows from equation 5.33:

$$Q^{\pi}(s, a = \pi'(s)) \geq Q^{\pi}(s, a) \text{ for all } a \in \mathcal{A} \tag{5.34}$$

$\pi'$ selected using equation 5.33 satisfies the condition of the policy improvement theorem in equation 5.28, with the proof sketched in equation 5.35. We have used $\sum_{a \in \mathcal{A}} \pi(a|s) = 1$ and equation 5.34:

$$\begin{aligned}
V^{\pi}(s) &= E_{a \sim \pi}\left[Q^{\pi}(s, a)\right] \\
&= \sum_{a \in \mathcal{A}} \pi(a|s) Q^{\pi}(s, a) \\
&\leq Q^{\pi}(s, a = \pi'(s)) \sum_{a \in \mathcal{A}} \pi(a|s) \\
&= Q^{\pi}(s, a = \pi'(s)) \\
&= Q^{\pi}(s, a = \pi'(s)) \text{ for all } s \in \mathcal{S}
\end{aligned} \tag{5.35}$$

Therefore, the policy improvement theorem implies that $\pi'$ is a better policy than $\pi$.

Generalized policy iteration involves starting with an initial policy (e.g., could be random), calculating the value function using policy evaluation, and using policy improvement to get a better policy, followed by repetition of the process of policy evaluation and improvement. This is repeated until we get to a stable policy. That policy is a local optimum. In finite state spaces, the process is guaranteed to converge because there are a finite number of policies and we are achieving monotonic improvement until we reach a local maximum. The generalized policy iteration algorithm is shown in algorithm listing 3.

---

**Algorithm 3**    Generalized Policy Iteration

**Require:**    Initial policy $\pi(s, a)$ and model of environment $p(s'|s, a)$ and $r(s', s, a)$

1: Initialize $\pi' \leftarrow \pi$.

2: **repeat** $\Delta \leftarrow 0$

3:     Evaluate the action value function for policy $\pi'$, $Q^{\pi'}(s,a)$.

4:     Using policy improvement on $Q^{\pi'}(s,a)$, formulate an improved policy $\pi^{imp}$.

5:     Evaluate the action value function for the improved policy, $Q^{\pi^{imp}}(s,a)$.

6:     $\Delta \leftarrow \max_{s \in S, a \in A} |Q^{\pi^{imp}}(s,a) - Q^{\pi'}(s,a)|)$

7:     $\pi' \leftarrow \pi^{imp}$

8: **until** $\Delta < \in$

---

When using generalized policy iteration to find an optimal policy, one must be cognizant of the exploration vs. exploitation trade-off. A greedy policy like the one shown in equation 5.33 relies on exploitation of values in the state-action value function, selecting the action corresponding to the highest state-action value function. However, for the generalized policy iteration algorithm where policy evaluation is followed by policy improvement, it is possible that certain actions in states have not been explored yet. A greedy policy will only exploit the actions that have been explored, ignoring the ones that have not been visited, thereby leading it to a suboptimal policy. To avoid this situation, exploitation is mixed with exploration in an $\epsilon$-greedy policy, shown in equation 5.36. This policy resorts to exploiting the known state-action value function with probability $1 - \epsilon$ and picks the action yielding the highest value, but also explores the selection of random action with probability $\epsilon$:

$$\pi^\epsilon(s) = \begin{cases} \text{argmax}_a Q(s,a) & \text{with probability } 1-\epsilon \\ random(a \in \mathcal{A}) & \text{with probability } \epsilon \end{cases} \tag{5.36}$$

Generalized policy iteration can be applied to the previous problem of finding the shortest path from entry to exit in a maze. The code is illustrated in Listing 5-6, and the optimal policy (shortest path) is shown in Figure 5-3.

***Listing 5-6.*** Solving the Maze Problem Using Generalized Policy Iteration

```
1    import numpy as np
2    from enum import Enum, unique
3    import logging
4    import time
5
6    logging.basicConfig(level = logging.INFO)
7    logger = logging.getLogger(__name__)
8
9    @unique
10   class Actions(Enum):
11       LEFT = 0
12       RIGHT = 1
13       UP = 2
14       DOWN = 3
15
16
```

```
17    class EpsGreedyPolicy(object):
18        def __init__(self, eps):
19            self.epsilon = eps
20            np.random.seed(50)
21
22        def nextAction(self, state, Q):
23            if np.random.random() < self.epsilon:
24                return Actions(np.random.choice(4))
25
26            maxval = MazeSolver.NEG_INFTY
27            act = None
28            for action in Actions:
29                if maxval < Q[state[0], state[1],
                    action.value]:
30                    maxval = Q[state[0], state[1],
                        action.value]
31                    act = action
32
33            if act is None:
34                act = action
35            return act
36
37
38    class MazeSolver(object):
39        NEG_INFTY = float(-1E10)
40        EPSILON = 0.1
41
42        def __init__(self, entry, exit):
43            self.gamma = 1.0
44            self.maxIter = 5000
45            self.mazeSize = (12, 12)
```

```
46          self.nActions = len(Actions)
47          self.entry = entry
48          self.exit = exit
49          self.walls =        {(5,1), (5,2), (5,3),
                                 (5,4), (5,5),
50                               (7,3), (7,4), (7,5), (7,6),
51                               (2,7), (3,7), (4,7), (5,7),
                                 (6,7), (7,7), (8,7),
52                               (4,9), (5,9), (6,9), (7,9),
                                 (8,9), (9,9), (10,9),
53                               (9,4), (10,4)}
54          self.actionMap = {Actions.LEFT : (-1, 0),
55                             Actions.RIGHT : (1, 0),
56                             Actions.UP : (0, -1),
57                             Actions.DOWN : (0, 1)}
58          # add bounding walls
59          for i in range(self.mazeSize[0]):
60              self.walls.add((i, 0))
61              self.walls.add((i, self.mazeSize[1]-1))
62
63          for j in range(self.mazeSize[1]):
64              self.walls.add((0, j))
65              self.walls.add((self.mazeSize[0]-1, j))
66          if self.entry in self.walls:
67              raise ValueError("Entry square is
                inadmissible")
68          if self.exit in self.walls:
69              raise ValueError("Exit square is
                inadmissible")
70          self.QStar = np.ndarray((self.mazeSize[0], self.
            mazeSize[1], self.nActions), dtype=np.float64)
```

```
71          self.initQStar(self.QStar)
72          self.policy = EpsGreedyPolicy(0.1)
73
74      def policyEvaluationAndImp(self):
75          # Using a greedy policy with updated Q implies an
            implicit policy improvement
76          itercount = 0
77          while itercount < self.maxIter:
78              itercount += 1
79              for i in range(self.mazeSize[0]):
80                  for j in range(self.mazeSize[1]):
81                      state = (i,j)
82                      if state in self.walls:
83                          continue
84                      action = self.policy.
                         nextAction(state, self.QStar)
85                      reward = self.
                         rewardFunc(state, action)
86                      nextstate = self.
                         transitionFunc(state, action)
87                      if nextstate not in self.walls:
88                          nextaction = self.policy.
                             nextAction(nextstate, self.QStar)
89                          nextq = self.QStar[nextstate[0],
                             nextstate[1], nextaction.value]
90                          newval = reward + self.
                             gamma * nextq
91                          self.QStar[state[0], state[1],
                             action.value] = newval
```

```
92                      else:
93                          self.QStar[state[0], state[1],
                                action.value] = reward
94
95      def transitionFunc(self, state0, action):
96          increments = self.actionMap[action]
97          return state0[0] + increments[0], state0[1] +
            increments[1]
98
99      def rewardFunc(self, state0, action):
100         state1 = self.transitionFunc(state0, action)
101         if state1 in self.walls:
102             return MazeSolver.NEG_INFTY
103         elif state1 == self.exit:
104             return 1
105         return -1
106
107     def initQStar(self, Q):
108         for i in range(self.mazeSize[0]):
109             for j in range(self.mazeSize[1]):
110                 square = (i,j)
111                 if square in self.walls:
112                     for action in Actions:
113                         Q[i, j, action.value] =
                            MazeSolver.NEG_INFTY
114                 else:
115                     for action in Actions:
116                         Q[i, j, action.value] = 0
117
118
```

```
119        def optPolicy(self):
120            # run generalized policy iteration
121            self.policyEvaluationAndImp()
122
123            optpath = [self.entry]
124            sq = self.entry
125            while sq != self.exit:
126                bestaction = self.policy.nextAction(sq,
                       self.QStar)
127                if bestaction is None:
128                    return optpath
129                sq = self.transitionFunc(sq, bestaction)
130                optpath.append(sq)
131
132            return optpath
133
134
135    if __name__ == "__main__":
136        entry = (1, 3)
137        exit = (1, 12)
138        maze_solver = MazeSolver(entry, exit)
139        path = maze_solver.optPolicy()
140        if path[-1] != exit:
141            logger.info("No path exists")
142
143        logger.info("->".join([str(p) for p in path]))
```

The output path can be seen in Listing 5-7.

***Listing 5-7.*** Computed Path Using Generalized Policy Iteration

```
1    (1, 3)->(1, 4)->(1, 5)->(1, 6)->(1, 7)->(1, 8)->(1, 9)->(1,
10)->(1, 11)->(1, 12)
```

# 5.3.3 Monte Carlo Method

The Monte Carlo method for estimating the value function is based on sampling episodes using a policy. This method can be used for cases where a model of the environment is not available but individual episodes are available. An episode is a sequence of state, action, reward, and next state tuples starting from an initial state and ending in a terminal state. It can also be used for cases where a model of the environment is available – in this case one can simulate experiences using the model.

There are two versions of Monte Carlo methods used in reinforcement learning – first-visit Monte Carlo and every-visit Monte Carlo. The first-visit Monte Carlo method considers the first time a state is visited in an episode, whereas the every-visit Monte Carlo method considers all visits in an episode. The pseudo-code for first-visit Monte Carlo is shown in algorithm listing 4.

Similarly, the pseudo-code for calculating the state-action value function $Q(s, a)$ using first-visit Monte Carlo is shown in algorithm listing 5. This algorithm uses the state value function calculated using algorithm listing 4.

---

**Algorithm 4**    Calculate the State Value Function Using the First-Visit Monte Carlo Method

**Require:**    Initial policy $\pi(s, a)$ and discount factor $\gamma$

1: $V(s) = 0$ for all $s \in S$.

2: $V'(s) \leftarrow$ empty list for all $s \in S$.

3: **repeat**

4:    Get an episode: a sequence of tuples $(s_t, a_t, r_t, s_{t+1})$.

5:    Initialize $R_t = 0$ for $t = 0, 1, ..., T - 1$.

6:   $R_{t-1} \leftarrow r_{T-1}$

7:   Create a dictionary $D$ to keep track of unique states within an episode. Add states in the episode to dictionary $D$, that is, $D \leftarrow s_0, s_1, \cdots, s_{T-1}$.

8:   **for** each $t \in [T-2, T-3, \cdots 0]$ **do**

9:       $R_t \leftarrow r_t + \gamma R_{t+1}$

10:   **end for**

11:   **for** each $s_t \in [s_0, s_1, \cdots, s_{T-1}]$ **do**

12:       **if** $s_t$ is in dictionary $D$ **then**

13:           Append $R_t$ to $V(s_t)$.

14:           Remove $s_t$ from dictionary $D$.

15:       **end if**

16:   **end for**

17: **until** All episodes have been processed

18: **for** each $s \in S$ **do**

19:     $V(s) \leftarrow$ average of rewards in list $V'(s_t)$.

20: **end for**

---

**Algorithm 5**    Calculate the State-Action Value Function Using the First-Visit Monte Carlo Method

**Require:**    Initial policy $\pi(s, a)$ and discount factor $\gamma$

1: $Q(s, a) = 0$ for all $s \in S$ and all $a \in A$.

2: $Q'(s, a) \leftarrow$ empty list for all $s \in S$ and all $a \in A$.

3: Calculate the value function $V(s)$ using the algorithm for computing the state value function and all episodes.

4: **repeat**

5:    Get an episode: a sequence of tuples $(s_t, a_t, r_t, s_{t+1})$.

6:    Create a dictionary $D$ to keep track of unique (state, action) tuples within an episode, $D \leftarrow (s_0, a_0), (s_1, a_1), \cdots (s_{T-1}, a_{T-1})$.

7:    **for** each $(s_t, a_t, r_t, s_{t+1})$ in the episode **do**

8:       **if** $(s_t, a_t)$ is in dictionary $D$ **then**

9:          Append $r_t + \gamma V(s_{t+1})$ to $Q'(s_t, a_t)$.

10:          Remove $(s_t, a_t)$ from dictionary $D$.

11:       **end if**

12:    **end for**

13: **until** All episodes have been processed

14: **for** each $s \in \mathcal{S}$ **do**

15:    **for** each $a \in \mathcal{A}$ **do**

16:       $Q(s, a) \leftarrow$ average of rewards in list $Q'(s, a)$.

17:    **end for**

18: **end for**

---

The asymptotic convergence rate for the first-visit Monte Carlo method is $\dfrac{1}{\sqrt{N}}$ where $N$ is the number of times a state is visited in all episodes (for first-visit Monte Carlo, a state cannot be visited more than once in each episode).

To illustrate an application of the first-visit Monte Carlo method, let us apply the method to evaluate the price of an American put option.

## Pricing an American Put Option

An American put option gives the holder the right but not the obligation to sell an underlying asset at the strike price at any time until the expiration of the option. For a non-dividend-paying stock, this could occur if the option is in the money at a certain time before expiration and the price is expected to go higher from that point onward.

Let us assume the stock price follows a geometric Brownian motion with constant volatility, given by equation 5.37. The solution of this equation is shown in equation 5.38. The put option matures in time $T$ and is written on an underlying asset (e.g., a stock) with constant volatility $\sigma$. Since the option can be hedged with the underlying asset, the rate of return, $\mu$, in the option pricing framework is the risk-free rate of return, $r_f$. Let us assume the term structure of risk-free rates to be flat, so that $r_f$ is a constant. We assume that the stock pays no dividend.

$$dS = \mu S dt + \sigma S dW_t \tag{5.37}$$

$$S_{t+\Delta t} = S_t \exp\left(\left(\mu - \frac{\sigma^2}{2}\right)\Delta t + \sigma\sqrt{\Delta t}\epsilon\right) \tag{5.38}$$

$$\epsilon \sim N(0,1)$$

The state space consists of the stock price and time. In this problem, we are guaranteed that no state will be visited twice in an episode because time only moves forward and it is a part of the state. At each time step, the stock price can move governed by equation 5.38 with $\mu = r_f$. Action space is discrete, with two actions at each step: exercise the option or not. Exercising the option terminates it and consequently the episode. Reward is the time-discounted value of final payoff. Hence, $\gamma = e^{-r_f \Delta t}$. Let us partition total time T into N equal partitions with $\Delta t = \dfrac{T}{N}$. With this choice of rewards and discount factor, the state value function $V(S, t)$ gives the present value of the option. The strike price of the option is $K$.

The optimum policy, $\pi^*$, in this example would be to exercise the option if its immediate exercise value is greater than the holding value of the option, or $\gamma V(S_{t+1}, t+1)$. This policy is shown in equation 5.39, with the boundary condition shown in equation 5.40. We generate Monte Carlo paths for the stock price using equation 5.38 and evaluate the state value function $V^{\pi^*}(S,t)$ using the first-visit Monte Carlo algorithm. Since the policy depends on the state value function, we must evaluate the policy on the path backward from time T down to 0. The boundary condition gives the value of $V^{\pi^*}(S_T,T)$ at expiration, T. Finally, $V^{\pi^*}(S_0,0)$ gives the value of the American option.

$$\pi^*(S_t,t) = \begin{cases} \text{Exercise,} & \text{if } K - S_t > \gamma V(S_{t+1}, t+1) \\ \text{Hold,} & \text{otherwise} \end{cases} \tag{5.39}$$

$$V^{\pi^*}(S_T,T) = \max(K - S_T, 0) \tag{5.40}$$

The full code is shown in Listing 5-8. Since we are only interested in $Q(s_0, t_0)$, the algorithm dispenses with keeping track of the full state-action value function. The state space in stock price is continuous, but we discretize it.

**Listing 5-8.** Valuation of an American Put Option Using the First-Visit Monte Carlo Method

```
1   import numpy as np
2   from enum import Enum, unique
3   import logging
4
5   logging.basicConfig(level = logging.INFO)
6   logger = logging.getLogger(__name__)
7
8   @unique
```

```
 9   class Actions(Enum):
10       HOLD = 0
11       EXERCISE = 1
12
13
14   class AmericanPutOption(object):
15
16       def __init__(self, S0, volat, strike, maturity, rf,
         time_steps=2000, npaths = 10000):
17           """
18           S0: initital stock price
19           volat: volatility of stock
20           strike: strike price
21           maturity: maturity of the option in years
22           rf: risk free rate (assumed constant) annual rate
23           time_steps: Number of time steps from 0 to
             maturity
24           """
25           self.nSamples = npaths
26           self.S0 = S0
27           self.volat = volat
28           self.K = strike
29           self.T = maturity
30           self.rf = rf
31           self.timeSteps = time_steps
32           self.gamma = np.exp(-rf/float(time_steps))
33           self.nPartSUp = time_steps*volat*np.sqrt(1.0/
             time_steps)
34           fac = volat*np.sqrt(1.0/time_steps)
35           self.probUp = (np.exp(rf/time_steps) -
             np.exp(-fac))/(np.exp(fac) - np.exp(-fac))
```

```
36
37      def generatePath(self):
38          path = [None] * (self.timeSteps + 1)
39          state = (0, np.log(self.S0))
40          path[0] = state
41          t = 0
42          incr = 1.0/self.timeSteps
43          stockval = np.log(self.S0)
44          incrS = self.volat*np.sqrt(1.0/self.timeSteps)
45          for i in range(self.timeSteps):
46              val = np.random.random()
47              if val <= self.probUp:
48                  stockval += incrS
49              else:
50                  stockval -= incrS
51              t += incr
52              path[i+1] = (t, stockval)
53          return path
54
55      def valueOnPath(self, path):
56          val = max(0, self.K - np.exp(path[-1][1]))
57          for i in range(len(path)-2, -1, -1):
58              exercise_val = self.K - np.exp(path[i][1])
59              val = max(self.gamma*val, exercise_val)
60          return val
61
62      def optionValue(self):
63          value = 0.0
64          for i in range(self.nSamples):
65              path = self.generatePath()
66              value += self.valueOnPath(path)
```

```
67            return value/self.nSamples
68
69
70   if __name__ == "__main__":
71       put_option = AmericanPutOption(20, 0.3, 21, 1, 0.005)
72       logger.info("Put option price: %f", put_option.
         optionValue())
```

Computer option price produced by the code is shown in Listing 5-9.

***Listing 5-9.*** Computed American Put Option Price

```
1   Put option price: 5.261670
```

# 5.3.4 Temporal Difference (TD) Learning

Temporal difference learning is an online algorithm for learning the value function of a policy using the experience of rewards. Like Monte Carlo methods, TD learning does not require a model of the environment. However, unlike Monte Carlo methods, it does not require a full episode of experience from the initial state to the terminal state in order to update the value function. It can use the observed reward value after a certain number of time steps, coupled with an existing estimate of the value function, to update the value function. In this sense, it is a bootstrapping method since it uses the current estimate of the value function to calculate an update to it. In practice, it is found to converge faster to the true value function than Monte Carlo methods primarily because it does not postpone learning until the end of the episode.

TD(0) learning uses observed reward and an estimate of the value function at the ensuing state to update the value function at the current state. The update rule for TD(0) learning is shown in equation 5.41. $\alpha$ is the learning rate and is typically chosen with a small value between 0 and

1 to ensure stability. TD(n) learning uses rewards from n states following the initial state along with the value function estimate at the last state to update the value function at the current state, as shown in equation 5.42. TD(n) learning has less bias but greater variance than TD(0) learning.

$$V^\pi(s) \leftarrow V^\pi(s) + \alpha \Big[ \big( r(s',s,a) + \gamma V^\pi(s') \big) - V^\pi(s) \Big] \tag{5.41}$$

$$V^\pi(s_0) \leftarrow V^\pi(s_0) + \alpha \Big[ \big( r(s_1,s_0,a_0) + \gamma r(s_2,s_1,a_1) + \ldots + \gamma^{n-1} r(s_{n+1},s_n,a_n) + \gamma^n V^\pi(s_{n+1}) \big) - V^\pi(s_0) \Big] \tag{5.42}$$

## SARSA

TD learning can be used to find an optimal policy starting with a nonoptimal policy using the SARSA algorithm, with the acronym SARSA standing for state, action, reward, and next state followed by a choice of action using an $\epsilon$-greedy policy shown in equation 5.36. The algorithm is shown in pseudo-code 6. It is essentially a combination of a TD update to the action value function followed by using an $\epsilon$-greedy policy and the updated action value function to find the next action.

## Valuation of an American Barrier Option

An American option can be exercised any time prior to or at maturity. By contrast, a European option can only be exercised at maturity. Other aspects of American barrier options are identical to their European counterparts, as described in a previous section. In this section, let us use SARSA to evaluate the fair market price of an American up-and-in barrier call option.

---

**Algorithm 6**    SARSA Algorithm for Finding the Optimal Policy

**Require:**    Discount factor $\gamma$, learning rate $\alpha$

1: $Q(s, a) = 0$ for all $s \in \mathcal{S}$ and all $a \in \mathcal{A}$ .

2: **repeat**

3:    **for** each $(s_t, a_t, r_t, s_{t+1})$ in an episode **do**

4:      $s \leftarrow s_t$

5:      Find action $a^*$ prescribed by an epsilon greedy policy using the state-action value function $Q(s, a)$ as shown in equation 5.43:

$$a^* = \begin{cases} \operatorname{argmax}_{a'} Q(s, a') \text{ with probability } 1 - \epsilon \\ random(a' \in \mathcal{A}) \text{ with probability } \epsilon \end{cases} \tag{5.43}$$

6:    **repeat**

7:        Take action $a^*$ and observe reward $r^*$ and next state $s^*$.

8:        Find action $a^{**}$ prescribed by an epsilon greedy policy using the state-action value function $Q(s^*, a')$ and equation 5.43 starting at state $s^*$.

9:        Update the action value function using
$Q(s, a^*) \leftarrow Q(s, a^*) + \alpha[(r^* + \gamma Q(s^*, a^{**})) - Q(s, a^*)]$.

10:      $s \leftarrow s^*, a^* \leftarrow a^{**}$

11:    **until** s is terminal state

12:  **end for**

13: **until** All episodes have been processed

---

As before, the state space consists of a two-dimensional grid with stock price along the X axis and time along the Y axis. Asset price moves up or down from a node, as shown in Figure 5-4. Let us assume the asset price follows log-normal dynamics shown in equation 5.18. There are two actions at each state: exercise the option or hold on to it. Let the state value function $V(S, t)$ denote the option price at state $(S, t)$. The state-action value function $Q(S, t, a_t)$ can be written using the state value function as shown in equation 5.44. $P_h(S, t)$ denotes the probability of reaching or exceeding barrier $B$ on a path to node $(S, t)$ and is computed using equation 5.25. $P_u$ and $P_d$ are the probabilities of moving to up and down nodes, respectively, from the current node, as given by equation 5.22. At a node where the option is exercised, its price is equal to $max(S_t - K, 0)$ conditional on the price hitting the barrier en route to that node. If the option is not exercised, its price is the expected discounted price at the next two nodes after taking barrier hitting probability into account, similar to equation 5.26 for a European option.

$$Q(S,t,a_t = \text{exercise}) = P_h(S,t)\max(S-K,0)$$

$$Q(S,t,a_t = \text{hold}) = e^{-r_f \Delta T} P_h(S,t)\left(P_u \frac{V(S_u,t+\Delta t)}{P_h(S_u,t+\Delta t)} + P_d \frac{V(S_d,t+\Delta t)}{P_h(S_d,t+\Delta t)}\right) \quad (5.44)$$

$$V(S,t) = \max\big(Q(S,t,a_t = \text{exercise}),Q(S,t,a_t = \text{hold})\big)$$

Using these definitions, let us price an American knock-in call option on a publicly traded stock with barrier $B$ of \$23, time to maturity $T$ of 2 months, risk-free rate $r_f$ of 0.5% per annum, volatility of stock $\sigma$ to be 20% per annum, moneyness or ratio of strike price to stock price $\dfrac{K}{S_0}$ to be 1.1, and current stock price $S_0$ to be \$20. This implies the strike price $K$ is \$22. These parameters are identical to the ones used to price the European knock-in call option.

The code for computing the option price using SARSA is shown in Listing 5-10. It uses a relatively high learning rate of 0.1, $\epsilon$ of 0.1 (for an $\epsilon$-greedy policy), and 2000 epochs. The calculated option price is \$0.005 and is higher than its European counterpart.

***Listing 5-10.*** Calculating an American Barrier Up-and-In Call Option's Price Using the SARSA Algorithm

```
1    import numpy as np
2    import logging
3    import matplotlib.pyplot as plt
4    from mpl_toolkits.mplot3d import Axes3D
5
6    logging.basicConfig(level=logging.DEBUG)
7    logger = logging.getLogger("root")
8
9
10   class AmericanKnockInCallOption(object):
11       def __init__(self, s0, strike, maturity, rf,
         volatility, barrier, minsteps=20,
12                       epsilon=0.1, epochs=2000, learning_
                         rate=0.1):
13           """
14           Initialize
15           :param s0: Initial price of underlying asset
16           :param strike: Strike price
17           :param maturity: Maturity in years
18           :param rf: Risk free rate (per annum)
19           :param volatility: expressed per annum
20           :param barrier: Barrier for this knock-in option
21           :param minsteps: Minimum number of time steps
```

```
22          :param epsilon: Epsilon defining the epsilon-
            greedy policy
23          :param epochs: Number of training epochs
24          :param learning_rate: Rate of learning
25          """
26          self.s0 = s0
27          self.strike = strike
28          self.barrier = barrier
29          self.maturity = maturity
30          self.rf = rf
31          self.vol = volatility
32          self.minSteps = minsteps
33          self.epsilon = epsilon
34          self.nepoch = epochs
35          self.alpha = learning_rate
36
37          self.deltaT = min(self.calculateDeltaT(),
            maturity/minsteps)
38          self.df = np.exp(-rf * self.deltaT)
39          self.sqrtTime = np.sqrt(self.deltaT)
40          expected = np.exp((rf +
            volatility*volatility/2.0)*self.deltaT)
41          self.up = np.exp(volatility * self.sqrtTime)
42          self.down = np.exp(-volatility * self.sqrtTime)
43          self.pUp = (expected - self.down)/(self.up -
            self.down)
44          self.pDown = 1.0 - self.pUp
45          self.ntime = int(np.ceil(maturity / self.deltaT))
46          self.stateValFunc = np.zeros((2*self.ntime, self.
            ntime), dtype=np.float32)
```

```
47              self.actionStateValFunc = np.zeros((2*self.ntime,
                self.ntime, 2), dtype=np.float32)
48              self.price = None
49              self.hitProb = self.calcBarrierHitProb()
50
51          def calcBarrierHitProb(self):
52              # calculate probability for t=0
53              hitprob = np.zeros((2*self.ntime, self.ntime),
                dtype=np.float32)
54              price = np.full(self.ntime*2, self.up, dtype=np.
                float32)
55              price[0] = self.s0 * (self.down ** self.ntime)
56              price = np.cumprod(price)
57              self.price = price
58
59              hitprob[:, -1] = np.where(price >= self.barrier,
                1.0, 0.0)
60
61              # for t = 1, 2, ... ntime-1
62              for j in range(self.ntime-2, -1, -1):
63                  for i in range(self.ntime-j, self.ntime+j+1):
64                      if price[i] >= self.barrier:
65                          hitprob[i, j] = 1.0
66                      else:
67                          hitprob[i, j] = self.pUp *
                            hitprob[i+1, j+1] + self.pDown *
                            hitprob[i-1, j+1]
68              return hitprob
69
70          def maturityCondition(self):
71              # values at time T
```

```
72          valFunc = self.stateValFunc
73          val = self.s0 * np.exp(-volatility * self.
            sqrtTime * self.ntime)
74          for i in range(2*self.ntime):
75              valFunc[i, -1] = self.hitProb[i, -1] *
                max(val - self.strike, 0)
76              val *= self.up
77
78      def epsilonGreedy(self, priceIndex, timeIndex):
79          if np.random.random() < self.epsilon:
80              return np.random.choice(2)
81          return np.argmax(self.
            actionStateValFunc[priceIndex, timeIndex, :])
82
83      def sarsaIters(self):
84          self.maturityCondition()
85          # generate episodes, perform epsilon greedy step
86          qFunc = self.actionStateValFunc
87          vFunc = self.stateValFunc
88          for iter in range(self.nepoch):
89              for j in range(self.ntime - 1, 0, -1):
90                  for i in range(self.ntime - j, self.ntime
                    + j, 1):
91                      # action = 0 -> Hold
92                      # action = 1 -> Exercise
93                      action = self.epsilonGreedy(i, j)
94                      if action == 0:
95                          val1 = 0
96                          if self.hitProb[i + 1, j] > 0:
97                              val1 = vFunc[i + 1, j] /
                                self.hitProb[i + 1, j]
```

```
98                         val2 = 0
99                         if self.hitProb[i - 1, j] > 0:
100                            val2 = vFunc[i - 1, j] /
                               self.hitProb[i - 1, j]
101                        newval = self.df * self.
                           hitProb[i, j - 1] * (self.pUp *
                           val1 + self.pDown * val2)
102                     else:
103                        newval = self.hitProb[i, j -
                           1] * max(self.price[i] - self.
                           strike, 0)
104                     qFunc[i, j - 1, action] += self.
                        alpha * (newval - vFunc[i, j-1]) #
                        SARSA update
105                     vFunc[i, j-1] = np.max(qFunc[i,
                        j - 1, :])
106
107         return vFunc[self.ntime, 0]
108
109     def calculateDeltaT(self):
110         val = self.vol / (self.rf + self.vol*self.
            vol/2.0)
111         return val*val
112
113     def plotPrice(self):
114         price = np.full(self.ntime * 2, self.up,
            dtype=np.float32)
115         price[0] = self.s0 * (self.down ** self.ntime)
116         price = np.cumprod(price)
117         time = np.full(self.ntime, self.deltaT, dtype=np.
            float32)
```

```
118            time[0] = 0
119            time = np.cumsum(time)
120            x, y = np.meshgrid(price, time)
121            fig = plt.figure()
122            axs = fig.add_subplot(111, projection='3d')
123            axs.plot_surface(x.T, y.T, self.stateValFunc)
124            axs.set_xlabel('Stock Price')
125            axs.set_ylabel('Time (Yrs)')
126            axs.set_zlabel('Option Price')
127            plt.show()
128
129            fig, axs = plt.subplots(1, 1, constrained_
               layout=True)
130            cs = axs.contourf(x.T, y.T, self.stateValFunc)
131            fig.colorbar(cs, ax=axs, shrink=0.85)
132            axs.set_title("American Barrier Knock-In Call
               Option")
133            axs.set_ylabel("Time to Maturity (yrs)")
134            axs.set_xlabel("Initial Stock Price")
135            axs.locator_params(nbins=5)
136            axs.clabel(cs, fmt="%1.1f", inline=True,
               fontsize=10, colors='w')
137            plt.show()
138
139
140    if __name__ == "__main__":
141        price = 20.0
142        strike = 22.0
143        maturity = 2.0/12.0
144        barrier = 23.0
145        volatility = 0.2
```

```
146        rf = 0.005
147        eoption = AmericanKnockInCallOption(price, strike,
           maturity, rf, volatility, barrier, minsteps=25)
148        simPrice = eoption.sarsaIters()
149        logger.info("simulated price: %f", simPrice)
150        eoption.plotPrice()
```

Computed option price produced by the code is shown in Listing 5-11.

***Listing 5-11.*** American Barrier Option Price Output from the Code

```
1   simulated price: 0.004923
```

The option price surface is shown in Figure 5-7 against stock price and time to maturity (in years). Contour plot of the option price is shown in Figure 5-8. As seen in the plots, the price of the knock-in option is close to 0 near option maturity and below the barrier at $23.

## Least Squares Temporal Difference (LSTD)

The least squares temporal difference algorithm combines temporal difference learning with gradient descent search to update the parameters of a value function. Like TD learning, it is an on-policy method. However, unlike TD learning, it is an offline algorithm because it requires data from all episodes to be available in order to perform a least squares minimization of the loss function. The algorithm was first proposed by Steven Bradtke and Andrew Barto in 1996.

The LSTD algorithm requires a parameterization of the value function. In prior examples, the state value function $V(s)$ has been represented as a table, with one value for each state. In most applied problems, the size of the state space renders explicit storage of the value function for each state intractable. For problems with continuous state space, storing explicit values for each state is clearly infeasible. Let us represent the value function using a parametric function, as shown in equation 5.45. $\phi(s)$

denotes a function of state written as a column vector, and $\theta$ denotes a column vector of parameters. $'$ denotes the transpose of a column vector.

$$V(s) = \phi(s)' \theta \qquad (5.45)$$

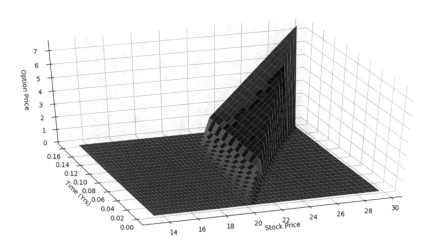

**Figure 5-7.**  *Option Price Surface for an American Barrier Up-and-In Call Option*

The algorithm learns the value of parameters by minimizing the mean square error loss function shown in equation 5.46 using the TD(0) target shown in equation 5.47:

$$\min_{\theta} \frac{1}{T} \sum_{t=1}^{T} \left( E_{s_{t+1}, a_t \sim \pi} \left[ r(s_{t+1}, s_t, a_t) + \gamma V(s_{t+1}) \right] - V(s_t) \right)^2$$

$$\min_{\theta} \frac{1}{T} \sum_{t=1}^{T} \left( \sum_{a_t} \pi(a_t | s_t) \sum_{s_{t+1}} p(s_{t+1} | s_t, a_t) \left[ r(s_{t+1}, s_t, a_t) + \gamma V(s_{t+1}) \right] - V(s_t) \right)^2 \qquad (5.46)$$

$$V(s_t)^{\pi}_{target} = E_{a_t \sim \pi, s_{t+1}} \left[ r(s_{t+1}, s_t, a_t) + \gamma V^{\pi}(s_{t+1}) \right] \qquad (5.47)$$

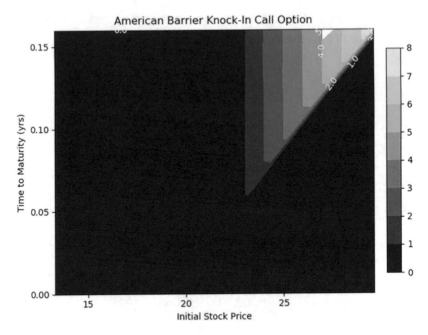

**Figure 5-8.** *Option Price Contour Plot for an American Barrier Up-and-In Call Option*

Using Monte Carlo draws from policy $\pi$, the loss function can be written as shown in equation 5.48. Substituting the parametric expression for the value function from equation 5.45 into the loss function, we can rewrite the loss function as shown in equation 5.49:

$$\min_{\theta} \frac{1}{T} \sum_{t=1}^{T} \left( r\left(s_{t+1}, s_t, a_t\right) + \gamma V\left(s_{t+1}\right) - V\left(s_t\right) \right)^2 \tag{5.48}$$

$$\min_{\theta} \frac{1}{T} \sum_{t=1}^{T} \left( r\left(s_{t+1}, s_t, a_t\right) + \gamma \phi\left(s_{t+1}\right)' \theta - \phi\left(s_t\right)' \theta \right)^2 \tag{5.49}$$

Differentiating the parameterized loss function in equation 5.49 with respect to parameters $\theta$ yields a system of equations 5.50 that can be solved using ordinary least squares (OLS) regression:

$$\frac{1}{T}\sum_{t=1}^{T}\left(r\left(s_{t+1},s_t,a_t\right)+\gamma\phi\left(s_{t+1}\right)'\theta-\phi\left(s_t\right)'\theta\right)\left(\gamma\phi\left(s_{t+1}\right)-\phi\left(s_t\right)\right)=0$$

$$\frac{1}{T}\sum_{t=1}^{T}r\left(s_{t+1},s_t,a_t\right)\left(\gamma\phi\left(s_{t+1}\right)-\phi\left(s_t\right)\right)+$$

(5.50)

$$\frac{1}{T}\sum_{t=1}^{T}\left(\gamma\phi\left(s_{t+1}\right)-\phi\left(s_t\right)\right)'\left(\gamma\phi\left(s_{t+1}\right)-\phi\left(s_t\right)\right)\theta=0$$

Equation 5.50 can be recast as shown in equation 5.51 to show more clearly that it is a system of ordinary linear regression equations $A\theta - b = 0$ with $A$ and $b$ shown in equation 5.51:

$$A\theta - b = 0$$

$$A = \frac{1}{T}\sum_{t=1}^{T}\left(\gamma\phi\left(s_{t+1}\right)-\phi\left(s_t\right)\right)'\left(\gamma\phi\left(s_{t+1}\right)-\phi\left(s_t\right)\right)$$

(5.51)

$$b = -\frac{1}{T}\sum_{t=1}^{T}r\left(s_{t+1},s_t,a_t\right)\left(\gamma\phi\left(s_{t+1}\right)-\phi\left(s_t\right)\right)$$

LSTD can also be written as a gradient descent rule as shown in equation 5.52. Here, $\theta^{prev}$ refers to the value of $\theta$ in the previous iteration:

$$\theta = \theta^{prev} + \alpha\frac{1}{T}\sum_{t=1}^{T}\left(r\left(s_{t+1},s_t,a_t\right)+\gamma\phi\left(s_{t+1}\right)'\theta^{prev}-\phi\left(s_t\right)'\theta^{prev}\right)\phi\left(s_t\right)$$ (5.52)

## Least Squares Policy Evaluation (LSPE)

The LSPE algorithm evaluates the state-action value function for a specified policy using the least squares method. Like LSTD, LSPE is an on-policy and offline algorithm. It uses a parameterization of the state-action value function shown in equation 5.53. $\phi_j(s, a)$ is an assumed function, and $\theta_j$ represents parameters. The parameterization is a practical approach to representing the state-action value function because explicit tabular representation of the state-action value function by state-action pairs can quickly become unwieldy with increasing size of state and action spaces.

$$Q(s,a) = \sum_{j=1}^{d} \theta_j \phi_j(s,a) = \phi(\boldsymbol{s},\boldsymbol{a})'\theta \qquad (5.53)$$

LSPE(0) uses TD(0) to calculate the target action value for a state, as shown in equation 5.54. If we had a tabular representation, equation 5.54 would give a set of $(\mathcal{S},\ \mathcal{A})$ equations in $(\mathcal{S},\ \mathcal{A})$ unknowns, giving a unique solution to the system. Plugging the parameterization in equation 5.53 into equation 5.54, we get the system of equations shown in equation 5.53. This equation has $d$ free variables in $\boldsymbol{w}$ and can be solved using the least squares method.

$$Q_{target}^{\pi}(s,a) = E_{s'}\left[ r(s',s,a) + \gamma \sum_{a'} \pi(a'|s')Q^{\pi}(s',a') \right]$$
$$= \sum_{s'} p(s'|s,a)\left[ r(s',s,a) + \gamma \sum_{a'} \pi(a'|s')Q^{\pi}(s',a') \right] \qquad (5.54)$$

$$w'\phi(\boldsymbol{s},\boldsymbol{a}) = \sum_{s'} p(s'|s,a)r(s',s,a) + \gamma \sum_{s} p(s'|s,a) \sum_{a'} \pi(a'|s')w'\phi(\boldsymbol{s'},\boldsymbol{a'})$$

$$\left[ \phi(\boldsymbol{s},\boldsymbol{a}) - \gamma \sum_{s} p(s'|s,a) \sum_{a'} \pi(a'|s')\phi(\boldsymbol{s'},\boldsymbol{a'}) \right]' w = \sum_{s'} p(s'|s,a)r(s',s,a) \qquad (5.55)$$

$$Aw = b$$

$$w = \left[A'A\right]^{-1} A'b$$

As was the case with LSTD, these expressions can be simplified if Monte Carlo draws are used.

# Least Squares Policy Iteration (LSPI)

The least squares policy iteration algorithm uses LSPE followed by $\epsilon$-greedy policy improvement in an iterative cycle until an optimum policy is attained. This is the same framework described in generalized policy iteration.

# Q-Learning

Q-learning is an off-policy, online learning algorithm that learns the state-action value function of an optimal policy. The algorithm selects the next action, $a'$, using an $\epsilon$-greedy policy in equation 5.36 using the update shown in equation 5.56 for a one-step update:

$$Q^*(s,a) \leftarrow Q^*(s,a) + \alpha\left[r(s',s,a) + \gamma Q^*(s',a') - Q^*(s,a)\right]$$

$$a' = \begin{cases} \text{argmax}_{\tilde{a}} Q^*(s',\tilde{a}) & \text{with probability } 1-\epsilon \\ random(\tilde{a} \in \mathcal{A}) & \text{with probability } \epsilon \end{cases} \qquad (5.56)$$

An n-step Q-learning update is shown in equation 5.57. $Q^*$ denotes the state-action value function for the optimal policy. An $\epsilon$-greedy policy is used at each of the n steps to determine the action. If following a stochastic update rule, sampled actions can be used for $a_1$, $a_2$, ... $a_N$ with the last action determined using an $\epsilon$-greedy policy.

$$Q^*(s,a) \leftarrow Q^*(s,a) + \alpha\left[r(s_1,s,a) + \gamma r(s_2,s_1,a_1) + \gamma^2 r(s_3,s_2,a_2) + \cdots\right.$$
$$\left. + \gamma^N Q^*(s_{N+1},a') - Q^*(s,a)\right]$$

$$a_1 = \begin{cases} \text{argmax}_{\tilde{a}} Q^*(s_1,\tilde{a}) & \text{with probability } 1-\epsilon \\ random(\tilde{a} \in \mathcal{A}) & \text{with probability } \epsilon \end{cases} \qquad (5.57)$$

...

$$a' = \begin{cases} \text{argmax}_{\tilde{a}} Q^*(s_{N+1},\tilde{a}) & \text{with probability } 1-\epsilon \\ random(\tilde{a} \in \mathcal{A}) & \text{with probability } \epsilon \end{cases}$$

Q-learning performs an update shown in equation 5.56 for each time step of an episode following the $\epsilon$-greedy policy until the terminal state is reached. Hence, it is off-policy because it does not use a model of the underlying policy to determine the next action. It only requires the initial state. The Q-learning algorithm is sketched in pseudo-code 7.

## Double Q-Learning

The Q-learning algorithm is plagued by two shortcomings: the updates performed are correlated in state space, and the action value updates are prone to overshooting beyond the optimal value. Correlation in state space arises because we are performing an update on states visited during an episode. Q-learning updates $Q^*(s_t, a^*)$ first, thereupon following the $\epsilon$-greedy policy until hitting the terminal state. The states visited are based upon knowledge of an existing $Q^*$ function, but this function has only been updated for states visited during an episode. This gives rise to the problem of serial correlation – exploration of correlated states followed by exploitation of existing $Q^*$ function values.

---

**Algorithm 7    Q-Learning for Finding the State-Action Value Function for an Optimal Policy**

**Require:**    Discount factor $\gamma$, learning rate $\alpha$

1: $Q(s, a) = 0$ for all $s \in S$ and all $a \in A$.

2: **repeat**

3:    **for** each ($s_t$, $\cdots$) in an episode **do**

4:      $s \leftarrow s_t$

5:      **repeat**

6:          Get the action $a^*$ prescribed by the $\epsilon$-greedy policy in state $s$.

7:          Take action $a^*$ and observe reward $r^*$ and next state $s^*$.

8:    $Q^*(s,a^*) \leftarrow Q^*(s,a^*) + \alpha \left[ r(s^*,s,a^*) + \gamma \max_{\tilde{a} \in A} Q(s^*,\tilde{a}) - Q^*(s,a^*) \right]$

9:              $s \leftarrow s^*$

10:      **until** s is terminal state

11:    **end for**

12: **until** All episodes have been processed

---

The problem associated with overshooting or overestimation of the correction term arises because the same $Q^*$ function is used for finding the optimal action and for evaluating the state-action value function. This can be seen more clearly from the update equation used in Q-learning (equation 5.56) rewritten as shown in equation 5.58, assuming the $\epsilon$-greedy policy has selected the condition with probability $1 - \epsilon$ as will usually be the case for small $\epsilon$. If $Q^*(s', a)$ is high for some action $a$, it will cause overshoot in corrected values for all $Q^*(s, a)$ where $s'$ state follows $s$.

$$Q^*(s,a) \leftarrow Q^*(s,a) + \alpha \left[ r(s',s,a) + \gamma Q^*(s', \operatorname*{argmax}_{\tilde{a} \in A} Q^*(s',\tilde{a})) - Q^*(s,a) \right] \quad (5.58)$$

Double Q-learning addresses the problem of correlation by storing the individual transitions from each episode in a replay buffer, $\mathcal{R}$, and selecting a random mini-batch of state transitions from this buffer for Q-learning. As Q-learning proceeds and the algorithm uncovers new transitions from states, each transition is added back to the replay buffer. The default replay buffer has a fixed capacity and drops oldest transitions to make room for new ones. There are several versions of the replay buffer in use: some prioritized by advantage functions and others by rank. These flavors of Q-learning will be discussed more thoroughly in the next chapter that delves into individual algorithms.

The problem of correction overestimation is addressed by using two Q functions: one Q function is used to calculate the target value – called the target value function $Q^*_{target}$ – and the other Q function is used to evaluate the optimum action, called the learned action value function $Q^*_{learned}$. Updates are applied at each iteration to the learned action value function, while the target action value function is updated periodically by copying the learned action value function. This is shown in equation 5.59. Even if $Q^*_{learned}$ overshoots for a certain state-action combination, those values are not propagated in updates because the target value function is used to calculate the value of the target and that function only changes to the learned action value function $Q^*_{learned}$ with a delay.

$$Q^*_{learned}(s,a) \leftarrow Q^*_{learned}(s,a) + \alpha \Big[ r(s',s,a) + \gamma Q^*_{target}(s', \text{argmax}_{\tilde{a} \in A} Q^*_{learned}(s',\tilde{a})) - Q^*_{learned}(s,a) \Big]$$

(5.59)

# Eligibility Trace

Eligibility trace is a mechanism for implementing N-step TD learning efficiently. One-step TD learning (equation 5.41) expands one state, using the existing value of the value function at the next state. N-step TD learning (equation 5.42) gives faster convergence because it can assign credit for a move N steps into the episode. As an example, if a high (or low) reward is earned after following N steps, TD(0) learning will have to wait until the value function of the next state has been updated with this information. This process is going to require several iterations because TD(0) learning propagates updated values one time step at a time. Equation 5.60 illustrates this numerically by writing the error from a TD(n) update. It can be seen that the TD(n) update reduces the error at each state by $\gamma^n$ for each update:

$$V_{new}^{\pi}(s_0) \leftarrow V_{old}^{\pi}(s_0) + \alpha \Big[ \big( r(s_1,s_0,a_0) + \gamma r(s_2,s_1,a_1) + \ldots +$$
$$\gamma^{n-1} r(s_{n+1},s_n,a_n) + \gamma^n V_{old}^{\pi}(s_{n+1}) \big) - V_{old}^{\pi}(s_0) \Big]$$

$$V_{old}^{\pi}(s_0) = E \Big[ r(s_1,s_0,a_0) + \gamma r(s_2,s_1,a_1) + \ldots + \gamma^{n-1} r(s_{n+1},s_n,a_n) + \gamma^n V_{old}^{\pi}(s_{n+1}) \Big]$$

$$V_{new}^{\pi}(s_0) - V_{old}^{\pi}(s_0) = \alpha \gamma^n \Big( V_{old}^{\pi}(s_{n+1}) - E \big[ V_{old}^{\pi}(s_{n+1}) \big] \Big)$$

(5.60)

Explicit unrolling of a policy for N steps at each state is inefficient due to the duplication of effort involved once we are in a specific state. Eligibility trace is a mechanism to implement N-step updates more efficiently. In order to derive the updates required for implementing TD(n) using eligibility trace, let us consider a constant $\lambda$ with the property $0 < \lambda < 1$. Let $G_t^{t+n}\big(V^{\pi}(s_t)\big)$ denote the n-step expansion of the value function, as shown in equation 5.61. This is the target used by the TD(n) algorithm. We want to write an expression for $V(s_0)$ using different n-step target values, assuming the chain of states has infinite length. This can be written as shown in equation 5.62, which corresponds to the expansion of the value function for all the states until the terminal state is reached:

$$G_t^{t+n}\big(V^{\pi}(s_{t+n+1})\big) = r(s_{t+1},s_t,a_t) + \gamma r(s_{t+2},s_{t+1},a_{t+1}) + \ldots +$$
$$\gamma^{n-1} r(s_{t+n+1},s_{t+n},a_{t+n}) + \gamma^n V^{\pi}(s_{t+n+1})$$

(5.61)

$$V^{\pi}(s_0) = (1-\lambda) \sum_{n=1}^{\infty} \lambda^{n-1} G_0^n \big(V^{\pi}(s_{n+1})\big)$$
$$= (1-\lambda) \Big[ G_0^1\big(V^{\pi}(s_2)\big) + \lambda^2 G_0^2\big(V^{\pi}(s_3)\big) + \ldots \Big]$$
$$= (1-\lambda) \Big[ r(s_1,s_0,a_0)\big(1 + \lambda + \lambda^2 + \ldots\big) + \lambda r(s_2,s_1,a_1)\big(1 + \lambda + \lambda^2 + \ldots\big) + \ldots \Big]$$
$$= r(s_1,s_0,a_0) + \lambda r(s_2,s_1,a_1) + \lambda^2 r(s_3,s_2,a_2) + \ldots$$
$$= G_0^{\infty}\big(V^{\pi}(s_{\infty})\big)$$

(5.62)

We want to write the TD(n) update rule in terms of TD(0) updates for different states. To do this, consider the correction in the update rule for TD($\infty$). Denoting the one-step TD correction at any state by $\delta s_t$ and using equations 5.63 and 5.62, we can write TD($\infty$) correction using one-step corrections as shown in equation 5.64:

$$\delta s_t = r\left(s_{t+1}, s_t, a_t\right) + V^\pi\left(s_{t+1}\right) - V^\pi\left(s_t\right) = TD^0_{correction}\left(s_t\right) \qquad (5.63)$$

$$
\begin{aligned}
TD^\infty_{correction} &= G_0^\infty\left(V^\pi\left(s_\infty\right)\right) - V^\pi\left(s_0\right)\\
&= r(s_1, s_0, a_0) + \lambda r(s_2, s_1, a_1) + \lambda^2 r(s_3, s_2, a_2) + \dots - V^\pi\left(s_0\right)\\
&= (1-\lambda)\sum_{n=1}^\infty \lambda^{n-1} G_0^n V^\pi\left(s_{n+1}\right) - (1-\lambda)\sum_{n=1}^\infty \lambda^{n-1} V^\pi\left(s_0\right)\\
&= (1-\lambda)\sum_{n=1}^\infty \lambda^{n-1}\left[G_0^n\left(V^\pi\left(s_{n+1}\right)\right) - V^\pi\left(s_0\right)\right]\\
&= (1-\lambda)\left[G_0^1\left(V^\pi\left(s_2\right)\right) - V^\pi\left(s_0\right) + \lambda(G_0^2\left(V^\pi\left(s_3\right)\right) - V^\pi\left(S_0\right)) + \dots\right]\\
&= (1-\lambda)\left[r_0 + \gamma V^\pi\left(s_1\right) - V^\pi\left(s_0\right) + \lambda(r_0 + \gamma r_1 + \gamma^2 V^\pi\left(s_2\right) - (r_0 + \gamma V^\pi\left(S_1\right)) + \dots\right]\\
&= (1-\lambda)\left[\delta s_0 + \lambda(\gamma(r_1 + \gamma V^\pi\left(s_2\right) - \gamma V^\pi\left(s_1\right))) + \dots\right]\\
&= (1-\lambda)\left[\delta s_0 + \lambda\gamma\delta s_1 + \lambda^2\gamma^2\delta s_2 + \dots\right] \qquad (5.64)
\end{aligned}
$$

Since the update rule in equation 5.60 has a factor $\alpha$ outside the correction, we merge the factor $1 - \lambda$ from equation 5.64 with $\alpha$ to write the update rule as shown in equation 5.65:

$$V^\pi_{new}\left(s_t\right) - V^\pi_{old}\left(s_t\right) = \alpha\left[\delta s_0 + \lambda\gamma\delta s_1 + \lambda^2\gamma^2\delta s_2 + \cdots\right] \qquad (5.65)$$

Equation 5.65 is an explicit expression relating the TD($\infty$) update to individual TD(0) corrections at different states. It's implementation is illustrated in pseudo-code 8 and is referred to as eligibility trace.

Eligibility traces can also be utilized in the evaluation of the state-action value function for a policy, as sketched in pseudo-code 9.

Eligibility traces can also be adapted for algorithms involving optimization of action value functions such as Q-learning or SARSA. For such algorithms, eligibility trace of a (state, action) pair is multiplied by $\lambda\gamma$

if the action corresponds to the greedy action giving the maximum action value function at that state. For an $\epsilon$-greedy policy, a random action is selected with a probability $\epsilon$. The Q-learning algorithm using eligibility traces is shown in pseudo-code 10.

## 5.3.5 Cartpole Balancing

Cartpole balancing is a benchmark reinforcement learning problem that seeks to balance a pole mounted on a slider that can slide along a frictionless rod as shown in the schematic in Figure 5-9. There are two actions available: applying a unit force (1 Newton) to the right or left, denoted as 1 or –1, respectively. The slider must remain within specific bounds on the rod, and its angle from the vertical line must likewise remain within bounds. The rod's length, mass, initial angle, initial angular velocity, and initial position are known.

---

**Algorithm 8**   Evaluating the Value Function Using Eligibility Trace

**Require:**   Policy $\pi$, discount factor $\gamma$, factor $\lambda$, and learning rate $\alpha$

1: $V(s) = 0$ for all $s \in \mathcal{S}$.

2: **repeat**

3:    **for** each $(s_t, \cdots)$ in an episode **do**

4:       $s \leftarrow s_t$

5:       Initialize set $S_{visited}=$ to empty set.

6:       Initialize $E(s) = 0$ for all $s \in \mathcal{S}$.

7:       **repeat**

8:          Add $s$ to $S_{visited}$.

9:          Get the action $a$ prescribed by policy $\pi$ in state $s$.

10:        Take action $a$ and observe reward $r$ and next state $s'$.

11:        $\delta \leftarrow r + \gamma V(s') - V(s)$

12:

13:        $E(s) \leftarrow 1$

14:

15:        **for** each state $s_v \in S_{visited}$ **do**

16:            $V(s_v) \leftarrow V(s_v) + \alpha E(s_v)\delta$

17:            $E(s_v) \leftarrow \lambda\gamma E(s_v)\delta \; s \leftarrow s'$

18:        **end for**

19:        $s \leftarrow s'$

20:    **until** s is terminal state

21:    **end for**

22: **until** All episodes have been processed

---

**Algorithm 9**   Evaluating the State-Action Value Function Using Eligibility Trace

**Require:**   Policy $\pi$, discount factor $\gamma$, factor $\lambda$, and learning rate $\alpha$

1: $Q(s, a) = 0$ for all $s \in \mathcal{S}$ and $a \in \mathcal{A}$ .

2: **repeat**

3:    **for** each $(s_t, \cdots)$ in an episode **do**

4:        $s \leftarrow s_t$

5:        Get the action $a$ prescribed by policy $\pi$ in state $s$.

6:      Initialize set $S_{visited}$= to empty set.

7:      Initialize $E(s, a) = 0$ for all $s \in S$ and $a \in \mathcal{A}$.

8:    **repeat**

9:          Take action $a$ and observe reward $r$ and next state $s'$.

10:      Add $(s, a)$ to $S_{visited}$.

11:        $\delta \leftarrow r + \gamma Q(s', a' = \pi(s')) - Q(s, a)$

12:

13:      $E(s, a) \leftarrow 1$

14:

15:      **for** each state $(s_v, a_v) \in S_{visited}$ **do**

16:          $Q(s_v, a_v) \leftarrow Q(s_v, a_v) + \alpha E(s_v, a_v)\delta$

17:          $E(s_v, a_v) \leftarrow \lambda\gamma E(s_v, a_v)\delta$

18:      **end for**

19:      $s \leftarrow s', a \leftarrow a'$

20:    **until** $s$ is terminal state

21:  **end for**

22: **until** All episodes have been processed

---

**Algorithm 10**    Using Eligibility Trace in Q-Learning

**Require:**    Discount factor $\gamma$, factor $\lambda$, and learning rate $\alpha$

1: $Q(s, a) = 0$ for all $s \in S$ and $a \in \mathcal{A}$.

2: **repeat**

3:   **for** each ($s_t, \cdots$) in an episode **do**

4:       $s \leftarrow s_t$

5:       Get the action $a'$ prescribed by an $\epsilon$-greedy policy in state $s$.

6:       Calculate the indicator variable $I_{a'} = 1$ if $\mathrm{argmax}_a Q(s,a) = a'$ and 0 otherwise.

7:       Initialize set $S_{visited}=$ to empty set.

8:       Initialize $E(s, a) = 0$ for all $s \in \mathcal{S}$ and $a \in \mathcal{A}$.

9:       **repeat**

10:          Take action $a'$ and observe reward $r$ and next state $s'$.

11:          Add $(s, a')$ pair to $S_{visited}$.

12:          $\delta \leftarrow r + \gamma Q(s', a') - Q(s, a)$

13:

14:          $E(s, a) \leftarrow 1$

15:

16:          **for** each state $(s_v, a_v) \in S_{visited}$ **do**

17:             $Q(s_v, a_v) \leftarrow Q(s_v, a_v) + \alpha E(s_v, a_v)\delta$

18:             **if** $I_{a'} == 1$ **then**

19:                $E(s_v, a_v) \leftarrow \lambda\gamma E(s_v, a_v)\delta$

20:             **else**

21:                $E(s_v, a_v) \leftarrow 0$

22:             **end if**

23:          **end for**

24:          $s \leftarrow s', a \leftarrow a'$

25:     **until** *s* is terminal state

26:   **end for**

27: **until** All episodes have been processed

---

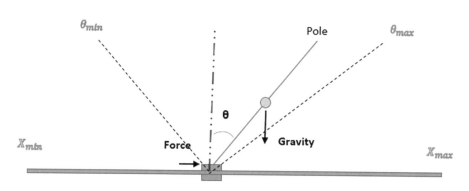

**Figure 5-9.** *Balancing a Cartpole by Applying a Series of Unit Forces on the Slider*

The motion of the slider and rod is governed by the laws of kinematics. There are four variables in kinematic laws describing this system – position and velocity of the slider and angle and angular velocity of the rod relative to the slider. The **aigym** Python library has an environment that simulates the cartpole's motion by implementing the laws of motion. Code in Listing 5-12 shows how to load the cartpole environment using **aigym** and examine its state space. The code shows that the slider can move between [-4.8, 4.8] and have vertical angle between [-4.19, 4.19] degrees. It can have any velocity and angular velocity.

**Listing 5-12.** Loading the Cartpole Environment in AIGym

```
1    import gym
2    env = gym.make("CartPole-v0")
3    print(env.observation_space)
```

```
4   Out[9]: Box([-4.8000002e+00 -3.4028235e+38 -4.1887903e-
    01 -3.4028235e+38], [4.8000002e+00 3.4028235e+38
    4.1887903e-01 3.4028235e+38], (4,), float32)
5
6   print(env.action_space)
7   Out[10]: Discrete(2)
```

The following observations will be helpful in understanding the code given in Listing 5-13:

1.  The environment gives a reward of 1 for each time step the pole remains upright and within the spatial and angular bounds. When the pole strays out of these bounds, the episode is terminated.

2.  The **AIGymEmulator** class encapsulates the cartpole environment provided by the **AIGym** library.

3.  The **Episode** class represents an episode as a collection of samples. Each sample (represented as an instance of class **Sample**) is a tuple containing state, action, reward, and next state.

4.  The state-action value function $Q(s, a)$ is represented using a neural network. Class **QNeuralNet** is used to represent the state-action value function. This class derives from the abstract base class **QFunction**.

5.  A three-layer sequential neural network is created using TensorFlow and passed to **QNeuralNet**.

6.  Since Q-learning is an off-policy method, it only uses the first state of an episode followed by an $\epsilon$-greedy policy derived from an existing state-action value network. Because of this feature, one-sample episodes are generated for training using initial slider position between –2 and 2, initial rod angle between –0.4 and 0.4 radians, initial velocity between –0.5 m/s and 0.5 m/s, and initial angular velocity between –0.5 rad/s and 0.5 rad/s.

7.  After training the agent, its performance is compared against a random agent that applies a random force on the slider. The length of an episode is restricted to 50 time steps. As seen in Table 5-1, the deep Q-network (DQN)-based agent outperforms the random agent. The table displays the total reward earned by each of the agents in 50 trials. The DQN agent reaches 50 time steps for all trials at which point the episode is terminated. Actual results obtained may differ slightly from those shown due to the use of random numbers in network weight initialization.

***Table 5-1.***  *Total Rewards Earned by the DQN Agent and Random Agent in 50 Trials with Maximum Episode Length of 50*

| DQN Agent | Random Agent |
|-----------|--------------|
| 50.0 | 26.0 |
| 50.0 | 15.0 |
| 50.0 | 28.0 |
| 50.0 | 44.0 |
| 50.0 | 16.0 |
| 50.0 | 13.0 |
| 50.0 | 18.0 |
| 50.0 | 27.0 |
| 50.0 | 27.0 |
| 50.0 | 20.0 |
| 50.0 | 41.0 |
| 50.0 | 35.0 |
| 50.0 | 8.0 |
| 50.0 | 44.0 |
| 50.0 | 23.0 |
| 50.0 | 15.0 |
| 50.0 | 14.0 |
| 50.0 | 10.0 |
| 50.0 | 12.0 |
| 50.0 | 50.0 |

(*continued*)

**Table 5-1.** (*continued*)

| DQN Agent | Random Agent |
|-----------|--------------|
| 50.0 | 13.0 |
| 50.0 | 23.0 |
| 50.0 | 22.0 |
| 50.0 | 12.0 |
| 50.0 | 22.0 |
| 50.0 | 13.0 |
| 50.0 | 17.0 |
| 50.0 | 14.0 |
| 50.0 | 16.0 |
| 50.0 | 39.0 |
| 50.0 | 36.0 |
| 50.0 | 50.0 |
| 50.0 | 20.0 |
| 50.0 | 13.0 |
| 50.0 | 20.0 |
| 50.0 | 20.0 |
| 50.0 | 19.0 |
| 50.0 | 20.0 |
| 50.0 | 27.0 |
| 50.0 | 17.0 |
| 50.0 | 15.0 |
| 50.0 | 12.0 |

(*continued*)

***Table 5-1.*** (*continued*)

| DQN Agent | Random Agent |
|-----------|--------------|
| 50.0 | 40.0 |
| 50.0 | 19.0 |
| 50.0 | 34.0 |
| 50.0 | 12.0 |
| 50.0 | 29.0 |
| 50.0 | 47.0 |
| 50.0 | 32.0 |
| 50.0 | 32.0 |

***Listing 5-13.*** Balancing a Cartpole Using a Deep Q-Network in Q-Learning

```
1    import time
2
3    import gym
4    import numpy as np
5    import pandas as pd
6    import tensorflow as tf
7
8    from src.learner.DQN import DQN
9    from src.lib.Emulator import AIGymEmulator
10   from src.lib.Episode import Episode
11   from src.lib.QFunction import QNeuralNet
12   from src.lib.Sample import Sample
13   from src.lib.ReplayBuffer import MostRecentReplayBuffer
14
15   tf.random.set_seed(10)
```

```
16    np.random.seed(10)
17
18
19    class CartpoleV0DQN(object):
20        def name(self):
21            name = self.__class__.__name__
22            if name.startswith("CartpoleV0"):
23                name = name[10:]
24            return name
25
26        def qfuncnetwork(self):
27            optimizer = tf.keras.optimizers.Adam()
28            #loss = tf.keras.losses.Huber(reduction=tf.keras.
                 losses.Reduction.SUM)
29            loss = tf.keras.losses.MeanSquaredError()
30            qnet = tf.keras.models.Sequential()
31            qnet.add(tf.keras.layers.Dense(10,
                 activation='relu', input_shape=(self.
                 nfeatures,)))
32            qnet.add(tf.keras.layers.Dense(50,
                 activation='relu'))
33            qnet.add(tf.keras.layers.Dense(self.nactions,
                 activation="linear"))
34            qnet.compile(optimizer=optimizer, loss=loss)
35            return qnet
36
37        def createAgent(self):
38            replay_buf = MostRecentReplayBuffer(2*self.
                 minibatchSize)
```

```
39          return DQN(self.qfunc, self.emulator,
            self.nfeatures, self.nactions, replay_
            buffer=replay_buf,
40                          discount_factor=self.
                            discountFactor, minibatch_
                            size=self.minibatchSize, epochs_
                            training=20)
41
42      def __init__(self):
43          self.nfeatures = 4
44          self.nactions = 2
45          self.testEpisodes = 50
46          self.maxTimeStepsInEpisode = 50
47          self.discountFactor = 1
48          self.minibatchSize = 20
49          qnet = self.qfuncnetwork()
50          self.qfunc = QNeuralNet(qnet, self.nfeatures,
            self.nactions)
51          # create emulator
52          self.envName = "CartPole-v0"
53          self.env = gym.make(self.envName)
54          self.emulator = AIGymEmulator(env_name=self.
            envName)
55          self.agent = self.createAgent()
56          self.train()
57
58      def generateTrainingEpisodes(self):
59          pos = np.arange(-2.0, 2.0, 4.0/10, dtype=np.
            float32)
60          vel = np.array([-0.5, 0, 0.5], dtype=np.float32)
```

```
61          angle = np.array([-0.4, -0.3, 0, 0.3, 0.4],
            dtype=np.float32)
62          angvel = np.array([-0.5, 0, 0.5], dtype=np.
            float32)
63          episodes = []
64          for p in pos:
65              for v in vel:
66                  for a in angle:
67                      for aa in angvel:
68                          state = np.array([p, v, a, aa],
                            dtype=np.float32)
69                          sample = Sample(state, 0,
                            1, None)
70                          episode = Episode([sample])
71                          episodes.append(episode)
72          return episodes
73
74      def train(self):
75          episodes = self.generateTrainingEpisodes()
76          return self.agent.fit(episodes)
77
78      def balance(self):
79          test_env = self.env
80          rewards = []
81          for i in range(self.testEpisodes):
82              obs0 = test_env.reset()
83              tot_reward = 0
84              fac = 1
85              for j in range(self.maxTimeStepsInEpisode):
86                  action, qval = self.agent.predict(obs0)
```

```
87              obs1, reward, done, info = test_env.
                step(action)
88              if done:
89                  break
90              tot_reward += fac * reward
91              obs0 = obs1
92              fac *= self.discountFactor
93          rewards.append(tot_reward)
94
95      random_agent_rewards = []
96      for i in range(self.testEpisodes):
97          test_env.reset()
98          tot_reward = 0
99          fac = 1
100         for j in range(self.maxTimeStepsInEpisode):
101             action = test_env.action_space.sample()
102             obs1, reward, done, info = test_env.
                step(action)
103             if done:
104                 break
105             tot_reward += fac * reward
106             fac *= self.discountFactor
107         random_agent_rewards.append(tot_reward)
108
109     result_df = pd.DataFrame({self.name(): rewards,
            "RandomAgent": random_agent_rewards})
110     print(result_df)
111     assert (np.mean(rewards) > np.mean(random_agent_
            rewards))
112
113
```

```
114    if __name__ == "__main__":
115        cartpole = CartpoleVoDQN()
116        cartpole.balance()
```

# 5.4  Policy Learning

Algorithms that optimize value functions (such as Q-learning) learn an optimal policy from the state-action value function $Q(s, a)$ using a greedy approach. The value function approach suffers from two disadvantages: a greedy approach produces a deterministic policy, and a small change in the action value function for an action can cause it to be selected or not selected, giving discontinuous jumps in action space. Many problems in reinforcement learning require a stochastic policy. An adversary can beat a deterministic policy with a knowledge of the policy. Policy learning learns a policy directly, without using a state-action value function, and does not suffer from the twin drawbacks for value function–based policy optimization. The foundational theory underpinning policy learning methods was formally introduced by Sutton et al. in a seminal paper published in 1999: "Policy Gradient Methods for Reinforcement Learning with Function Approximation." Let us look at the central concept of the paper – the policy gradient theorem.

## 5.4.1  Policy Gradient Theorem

Let the policy $\pi$ be parameterized by $\theta$, that is, let us denote the policy being learned by $\pi(a|s, \theta)$. The expected discounted sum of rewards earned by following policy $\pi$ beginning from state $s_0$ is given by $V^\pi(s_0)$, as shown in equation 5.66:

$$
\begin{aligned}
V^\pi(s_0) &= E_{a \sim \pi, s_1, s_2, \dots} \left[ \sum_{t=0}^{\infty} \gamma^t r(s_{t+1}, s_t, a_t) \right] \\
&= \sum_{a \in A} \pi(a|s_0, \theta) Q^\pi(s_0, a)
\end{aligned}
\tag{5.66}
$$

The policy gradient theorem expresses the derivative of the value function with respect to the policy function's parameters $\theta$. It is a foundational theorem because the derivative is used to update the policy function parameters $\theta$ using stochastic gradient ascent in order to maximize the expected discounted rewards or $V^\pi(s)$ as seen in equation 5.66. The policy gradient theorem is shown in equation 5.67. $d^\pi(s)$ denotes the discounted probability of landing in state $s$, as shown in equation 5.3.

$$\frac{\partial V^\pi(s_0)}{\partial \theta} = \sum_{s\in S} d^\pi(s|s_0) \sum_{a\in A} \frac{\partial \pi(a|s,\theta)}{\partial \theta} Q^\pi(s,a) \qquad (5.67)$$

In practice, $Q^\pi(s, a)$ in the policy gradient theorem 5.67 is often replaced by advantage, $A^\pi(s, a)$, from equation 5.11. We will see the benefit of working with the advantage function in the following sections. Equation 5.68 shows why the replacement of the action value function with the advantage function is an equivalent statement of the policy gradient theorem:

$$\sum_{s\in S} d^\pi(s|s_0) \sum_{a\in A} \frac{\partial \pi(a|s,\theta)}{\partial \theta} A^\pi(s,a)$$

$$= \sum_{s\in S} d^\pi(s|s_0) \sum_{a\in A} \frac{\partial \pi(a|s,\theta)}{\partial \theta} (Q^\pi(s,a) - V^\pi(s))$$

$$= \sum_{s\in S} d^\pi(s|s_0) \sum_{a\in A} \frac{\partial \pi(a|s,\theta)}{\partial \theta} (Q^\pi(s,a)$$

$$- \sum_{s\in S} d^\pi(s|s_0) \sum_{a\in A} \frac{\partial \pi(a|s,\theta)}{\partial \theta} V^\pi(s)$$

$$= \frac{\partial V^\pi(s_0)}{\partial \theta} - \sum_{s\in S} d^\pi(s|s_0) V^\pi(s) \frac{\partial \sum_{a\in A} \pi(a|s,\theta)}{\partial \theta}$$

$$= \frac{\partial V^\pi(s_0)}{\partial \theta} - \sum_{s\in S} d^\pi(s|s_0) V^\pi(s) \frac{\partial 1}{\partial \theta}$$

$$= \frac{\partial V^\pi(s_0)}{\partial \theta} \qquad (5.68)$$

The proof of the policy gradient theorem is sketched in equation 5.69:

$$\frac{\partial V^{\pi}(s_0)}{\partial \theta} = \sum_{a \in A} \frac{\partial \pi(a \mid s_0, \theta)}{\partial \theta} Q^{\pi}(s_0, a) + \sum_{a \in A} \pi(a \mid s_0, \theta) \frac{\partial Q^{\pi}(s_0, a)}{\partial \theta}$$

$$= \sum_{a \in A} \frac{\partial \pi(a \mid s_0, \theta)}{\partial \theta} Q^{\pi}(s_0, a) + \sum_{a \in A} \pi(a \mid s_0, \theta)$$

$$\frac{\partial (\sum_{s \in S} p(s \mid s_0, a)(r(s, s_0, a) + r \sum_{a' \in A} \pi(a' \mid s, \theta) Q^{\pi}(s, a')))}{\partial \theta}$$

$$= \sum_{a \in A} \frac{\partial \pi(a \mid s_0, \theta)}{\partial \theta} Q^{\pi}(s_0, a) + \gamma \sum_{a \in A} \sum_{s \in S} \pi(a \mid s_0, \theta) p(s \mid s_0, a)$$

$$\sum_{a' \in A} \frac{\partial \pi(a' \mid s, \theta)}{\theta} Q^{\pi}(s, a') + \ldots$$

$$= \sum_{s \in S} d^{\pi}(s \mid s_0) \sum_{a \in A} \frac{\partial \pi(a \mid s, \theta)}{\theta} Q^{\pi}(s, a) \tag{5.69}$$

Because policy learning does not explicitly learn the state-action value function $Q^{\pi}(s, a)$, this value needs to be estimated. Algorithms differ in their approach to estimating the state-action value function: some use a simple TD(0) expansion, while others such as actor-critic methods maintain a separate model for the value function that is learned concurrently with the policy.

## 5.4.2 REINFORCE Algorithm

The REINFORCE algorithm was proposed by R. J. Williams in 1992. It is a stochastic gradient descent algorithm that approximates the action value function $Q^{\pi}(s, a)$ in equation 5.67 using the rewards observed in a sample episode. Using a policy $\pi(a \mid s, \theta)$, it samples an episode and calculates the sum of expected discounted rewards $R^{\pi}(s_0, a_0)$ as a proxy for $Q^{\pi}(s_0, a_0)$ as shown in equation 5.70. The update rule used in the algorithm is shown in equation 5.71 and is performed once for each state in the episode. Factor $\frac{1}{\pi(a \mid s, \theta)}$ is required because of an implicit $\pi(a \mid s, \theta)$ introduced due to sampling using policy $\pi$.

$$R^{\pi}\left(s_0,a_0\right)=r\left(s_1,s_0,a_0\right)+\gamma r\left(s_2,s_1,a_1\right)+\gamma^2 r\left(s_3,s_2,a_2\right)+\ldots \qquad (5.70)$$

$$\theta_{new} \leftarrow \theta_{old} + \alpha\left(\frac{1}{\pi\left(a_0|s_0,\theta_{old}\right)}\frac{\partial\pi\left(a_0|s_0,\theta_{old}\right)}{\partial\theta}R^{\pi}\left(s_0,a_0\right)\right)$$

$$\Rightarrow \theta_{new} \leftarrow \theta_{old} + \alpha\left(\frac{\partial\log\pi\left(a_0|s_0,\theta_{old}\right)}{\partial\theta}R^{\pi}\left(s_0,a_0\right)\right) \qquad (5.71)$$

The complete REINFORCE algorithm is sketched in pseudo-code 11.

---

**Algorithm 11**    REINFORCE Algorithm for Policy Learning

**Require:**    Parameterized policy $\pi(a|s,\theta)$, discount factor $\gamma$, and learning rate $\alpha$

1: **repeat**

2:    Generate an episode by sampling from the environment.

3:    **for** each $(s_t, a_t, r_t, s_{t+1})$ in an episode **do**

4:        $s \leftarrow s_t$

5:        $a \leftarrow a_t$

6:        Start in a state $s$, take the action $a$ sampled from $\pi(a|s,\theta)$, transition to $s'$, and repeat until the terminal state is reached.

7:        $R^{\pi}(s,a) = r(s',s,a) + \gamma r(s'',s',a') + \gamma^2 r(s''',s'',a'') + \cdots$

8:        $\theta \leftarrow \theta + \alpha\left(\frac{1}{\pi(a|s,\theta)}\frac{\partial\pi(a|s,\theta)}{\partial\theta}R^{\pi}(s,a)\right)$

9:    **end for**

10: **until** All episodes have been processed

---

# 5.4.3 Policy Gradient with State-Action Value Function Approximation

The state-action value function $Q^\pi(s, a)$ is often represented as a parameterized function (e.g., using a neural network) in contrast to the REINFORCE method that uses the reward experience from an episode. According to the theorem of policy gradient with function approximation, we can replace $Q^\pi(s, a)$ in the policy gradient theorem with the learned approximation to the state-action value function as shown in equation 5.72:

$$\frac{\partial V^\pi(s_0)}{\partial \theta} = \sum_{s \in S} d^\pi(s|s_0) \sum_{a \in A} \frac{\partial \pi(a|,s|,\theta)}{\partial \theta} f(s,a,\mathbf{w}) \qquad (5.72)$$

In order to understand the theorem, let us denote the learned state-action value function as $f(s, a, \mathbf{w})$ where $\mathbf{w}$ denotes the vector of parameters of the state-action value function approximation. Let $\hat{Q}^\pi(s,a)$ denote an unbiased approximation to the state-action value function such as TD(0) target $r + \gamma \sum_{a' in A} \pi(a'|s') f(s',a',\mathbf{w})$ or the discounted sum of rewards from an episode $R^\pi(s, a)$ used by the REINFORCE algorithm. The parameters $\mathbf{w}$ are updated using a gradient descent rule shown in equation 5.73:

$$\min_{\mathbf{w}} \left( Q^\pi(s,a) - f(s,a,\mathbf{w}) \right)^2$$

$$\mathbf{w}_{new} \leftarrow \mathbf{w}_{old} + \alpha \left( Q^\pi(s,a) - f(s,a,\mathbf{w}) \right) \frac{\partial f(s,a,\mathbf{w})}{\partial \mathbf{w}} \qquad (5.73)$$

Once the updates to $\mathbf{w}$ have converged, condition 5.74 is satisfied:

$$\sum_{s \in S} d^\pi(s) \sum_{a \in A} \pi(s,a) \left( Q^\pi(s,a) - f(s,a,\mathbf{w}) \right) \frac{\partial f(s,a,\mathbf{w})}{\partial \mathbf{w}} = 0 \qquad (5.74)$$

The proof of this assertion is shown in equation 5.75 and uses the fact that $\hat{Q}^{\pi}(s,a)$ is an unbiased estimator of $Q^{\pi}(s,a)$:

$$\Delta \mathbf{w} = \alpha \left( \hat{Q}^{\pi}(s,a) - f(s,a,\mathbf{w}) \right) \frac{\partial f(s,a,\mathbf{w})}{\partial \mathbf{w}}$$

Upon convergence, $\Delta \mathbf{w} = 0$

$$\left( \hat{Q}^{\pi}(s,a) - f(s,a,\mathbf{w}) \right) \frac{\partial f(s,a,\mathbf{w})}{\partial \mathbf{w}} = 0 \forall s, \forall a \sim \pi(s,a) \tag{5.75}$$

$$\sum_{s \in S} d^{\pi}(s) \sum_{a \in A} \pi(s,a) \left( Q^{\pi}(s,a) - f(s,a,\mathbf{w}) \right) \frac{\partial f(s,a,\mathbf{w})}{\partial \mathbf{w}} = 0$$

The action value function approximator $f(s, a, \mathbf{w})$ is said to be compatible with policy parameterization if the condition in equation 5.76 is satisfied:

$$\frac{\partial f(s,a,\mathbf{w})}{\partial \mathbf{w}} = \frac{\partial \pi(s,a,\theta)}{\partial \theta} \frac{1}{\pi(s,a,\theta)} \tag{5.76}$$

If the action value function approximator is compatible with policy parameterization and we optimize the weights $\mathbf{w}$ of the value function approximator using equation 5.74 until convergence, the policy gradient theorem can be rewritten as shown in equation 5.72 by substituting equation 5.76 in equation 5.75.

$$\sum_{s\in\mathcal{S}} d^{\pi}(s) \sum_{a\in\mathcal{A}} \pi(s,a)(Q^{\pi}(s,a) - f(s,a,\mathbf{w})) \frac{\partial f(s,a,\mathbf{w})}{\partial \mathbf{w}} = 0$$

$$\Rightarrow \sum_{s\in\mathcal{S}} d^{\pi}(s) \sum_{a\in\mathcal{A}} \pi(s,a) Q^{\pi}(s,a) \frac{\partial f(s,a,\mathbf{w})}{\partial \mathbf{w}} =$$

$$\sum_{s\in\mathcal{S}} d^{\pi}(s) \sum_{a\in\mathcal{A}} \pi(s,a) f(s,a,\mathbf{w}) \frac{\partial f(s,a,\mathbf{w})}{\partial \mathbf{w}}$$

$$\Rightarrow \sum_{s\in\mathcal{S}} d^{\pi}(s) \sum_{a\in\mathcal{A}} \pi(s,a) Q^{\pi}(s,a) \frac{\partial \pi(s,a,\theta)}{\partial \theta} \frac{1}{\pi(s,a,\theta)} =$$

$$\sum_{s\in\mathcal{S}} d^{\pi}(s) \sum_{a\in\mathcal{A}} \pi(s,a) f(s,a,\mathbf{w}) \frac{\partial \pi(s,a,\theta)}{\partial \theta} \frac{1}{\pi(s,a,\theta)}$$

$$\Rightarrow \sum_{s\in\mathcal{S}} d^{\pi}(s) \sum_{a\in\mathcal{A}} Q^{\pi}(s,a) \frac{\partial \pi(s,a,\theta)}{\partial \theta} =$$

$$\sum_{s\in\mathcal{S}} d^{\pi}(s) \sum_{a\in\mathcal{A}} f(s,a,\mathbf{w}) \frac{\partial \pi(s,a,\theta)}{\partial \theta}$$

$$\Rightarrow \frac{\partial V^{\pi}(s_0)}{\partial \theta} = \sum_{s\in\mathcal{S}} d^{\pi}(s \mid s_0) \sum_{a\in\mathcal{A}} \frac{\partial \pi(a \mid s,\theta)}{\partial \theta} f(s,a,\mathbf{w}) \tag{5.77}$$

## 5.4.4 Policy Learning Using Cross Entropy

The cross entropy method can be used instead of gradient ascent in order to optimize a policy. An advantage of using the cross entropy method over gradient ascent is that it does not get trapped in local maxima. This benefit comes at a cost of greater time complexity. The cross entropy method seeks to iteratively improve a policy by sampling actions from a distribution centered around actions giving high rewards. The algorithm is sketched pseudo-code 12.

**Algorithm 12    Policy Optimization Using the Cross Entropy Method**

**Require**: State-action value function $Q(s, a)$ on continuous action space $\mathcal{A}$, state $s$, percentile K

1: $\mu \leftarrow 0, \sigma \leftarrow 1$

2: **repeat**

3:    $A \leftarrow a_i$ with $a_i$ sampled from Gaussian distribution $N(\mu, \sigma^2)$.

4:    Normalize action values $\tilde{A} \leftarrow \tilde{a}_i$ using $\tilde{a}_i = \tanh(a_i)$ so that $\tilde{a}_i$ lies between $-1$ and $1$.

5:    $\mathcal{Q} \leftarrow q_i : q_i = Q(s, \tilde{a}_i)$

6:    Sort $\mathcal{Q}$ in descending order of action values and keep the top K percentile of values $I \leftarrow sort(\mathcal{Q})_k, k \in [1, 2, \cdots, kN]$.

7:    $\mu \leftarrow \dfrac{1}{kN} \sum_{i \in I} a_i$

8:    $\sigma^2 \leftarrow var_{i \in I} a_i$

9:    Keep the action that gave the best value $q_i$, that is, the action corresponding to the first element of the sorted array $I$. $a' \leftarrow a_0$.

10: **until** $\mu$ and $\alpha$ have stopped changing

11: Return $\pi_{CEM}(s) = a'$.

# 5.5 Actor-Critic Algorithms

Policy learning methods address a major shortcoming of value function–based methods for reinforcement learning. Value function methods are susceptible to the curse of dimensionality: as the dimensions of state and action spaces grow, the number of state-action pairs grows exponentially. Policy learning methods circumvent this problem by learning the policy directly using policy gradient coupled with a parametric representation of the policy. This approach has one shortcoming: policy parameter updates prescribed by policy gradient in equation 5.67 display a high variance due to stochastic sampling of paths. In order to reduce the variance, a baseline function $b(s)$ is subtracted from $Q^\pi(s, a)$ to get the update rule shown in equation 5.78. Any function that depends only on state $s$ can be used as a permissible baseline $b(s)$. As before, $\theta$ denotes policy function parameters.

$$\theta \leftarrow \theta + \alpha \frac{\partial V^\pi(s_0)}{\partial \theta}$$

$$\theta \leftarrow \theta + \alpha \sum_{s \in S} d^\pi(s|s_0) \sum_{a \in A} \frac{\partial \pi(a|s,\theta)}{\partial \theta} \left(Q^\pi(s,a) - b(s)\right)$$

(5.78)

Subtracting a baseline function $b(s)$ is permissible because of equation 5.79:

$$\sum_{s \in S} d^\pi(s|s_0) \sum_{a \in A} \frac{\partial \pi(a|s,\theta)}{\partial \theta} b(s)$$

$$= \sum_{s \in S} d^\pi(s|s_0) b(s) \sum_{a \in A} \frac{\partial \pi(a|s,\theta)}{\partial \theta}$$

$$= \sum_{s \in S} d^\pi(s|s_0) b(s) \frac{\partial \sum_{a \in A} \pi(a|s,\theta)}{\partial \theta}$$

(5.79)

$$= \sum_{s \in S} d^\pi(s|s_0) b(s) \frac{\partial 1}{\partial \theta}$$

$$= 0$$

If the baseline is chosen to be the value function in that state $V^\pi(s)$, the parameter updates have minimum variance as shown in equation 5.80:

$$\text{Variance} = \sum_{s \in S} d^\pi(s \mid s_0) \sum_{a \in A} \pi(s,a)[Q^\pi(s,a) - b(s)]^2$$

$$\text{minimize variance} \Rightarrow \sum_{s \in S} d^\pi(s \mid s_0) \sum_{a \in A} \pi(s,a)[Q^\pi(s,a) - b(s)] = 0$$

$$\Rightarrow \sum_{s \in S} d^\pi(s \mid s_0) \sum_{a \in A} \pi(s,a)Q^\pi(s,a) = \sum_{s \in S} d^\pi(s \mid s_0) \sum_{a \in A} \pi(s,a)b(s)$$

$$\Rightarrow \sum_{s \in S} d^\pi(s \mid s_0) \sum_{a \in A} \pi(s,a)Q^\pi(s,a) = \sum_{s \in S} d^\pi(s \mid s_0)b(s) \sum_{a \in A} \pi(s,a)$$

$$\Rightarrow \sum_{s \in S} d^\pi(s \mid s_0)V^\pi(s) = \sum_{s \in S} d^\pi(s \mid s_0)b(s)$$

$$\Rightarrow V^\pi(s) = b(s)$$

$$(5.80)$$

Using the value function as a baseline poses a problem because it is not available while we are still learning the policy $\pi$. Actor-critic methods surmount this problem by keeping two agents: critic that is responsible for evaluating the current policy to update the value function and actor that updates the policy using the value function predicted by the critic as a baseline using equation 5.68. Actor and critic learning needs to proceed in lockstep because both are dependent on each other: actor relying on critic to provide an updated baseline (value function) and critic relying on actor to update the policy, which will be reflected in the updated value function.

To summarize, in actor-critic methods, the actor optimizes the policy using equation 5.81, which the critic uses to update the action value function by minimizing the sum of square deviations of the action value function from a given target, such as a TD target as shown in equation 5.82. The actor uses the value function as a baseline to enhance numerical stability of policy learning and depends on the critic to provide an estimate of the value function, while the critic depends upon the actor to provide the updated policy. The policy function used by the actor has been parameterized with $\theta$, and the value function used by the critic has been parameterized with $\nu$.

$$\max_{\pi} J^{\pi}(s_0) = \sum_{s \in S} d^{\pi}(s) \sum_{a \in A} \pi(a|s) \sum_{s' in S} p(s',s,a) r(s',s,a)$$

$$= \int_s d^{\pi}(s) \int_a \pi(a|s) \int_{s'} p(s',s,a) r(s',s,a) ds' da ds = V^{\pi}(s_0)$$

$$\max_{\theta} V^{\pi}(s_0)$$

(5.81)

$$\theta_{new} \leftarrow \theta_{old} + \alpha_a \frac{\partial V^{\pi}(s_0)}{\partial \theta}$$

$$\theta_{new} \leftarrow \theta_{old} + \alpha_a \sum_{s \in S} d^{\pi}(s|s_0) \sum_{a \in A} \frac{\partial \pi(a|s,\theta)}{\partial \theta} \left( Q^{\pi}(s,a) - V^{\pi}(s) \right)$$

$$\min_{v} E_{s',s,a \sim \pi} \left[ r(s',s,a) + \gamma V^{\pi}(s') - V^{\pi}(s) \right]^2$$

$$V^{\pi}(s) = V^{\pi}(s,v)$$

(5.82)

$$v_{new} \leftarrow v_{old} + \alpha_c \left[ r(s',s,a) + \gamma V^{\pi}(s') - V^{\pi}(s) \right] \frac{\partial V^{\pi}(s,v)}{\partial v}$$

## 5.5.1 Stochastic Gradient–Based Actor-Critic Algorithms

Stochastic gradient–based actor-critic algorithms take an episode and update actor and critic parameters based on stochastic gradient ascent/descent using the episode. This gives the following parameter update equations for the actor (equation 5.83) and critic (equation 5.84):

Policy $\pi(a|s,\theta)$

$$\theta_{new} \leftarrow \theta_{old} + \alpha_a \frac{\partial \log \pi(a|s,\theta)}{\partial \theta} \left( Q^{\pi}(s,a) - V^{\pi}(s) \right)$$

(5.83)

Value function $V^{\pi}(s) = V^{\pi}(s,v)$

$$v_{new} \leftarrow v_{old} + \alpha_c \left[ r(s',s,a) + \gamma V^{\pi}(s') - V^{\pi}(s) \right] \frac{\partial V^{\pi}(s,v)}{\partial v}$$

(5.84)

# 5.5.2 Building a Trading Strategy

In this section, let us apply an actor-critic algorithm to build a trading strategy for trading two stocks: Microsoft (MSFT) and Boeing (BA). The actor-critic strategy's performance is compared against the baseline buy-and-hold strategy. The portfolio holding period is 1 month – or 21 trading days. To make the comparison realistic, transaction cost of 0.1% of traded notional amount is added. Each day in the 21-day period, the actor-critic strategy has a chance to rebalance its two-stock portfolio, incurring transaction cost in the process.

Both strategies are given $2 worth of capital. On the first day, $1 is invested in each of the two stocks, assuming fractional shares. The buy-and-hold strategy holds this portfolio for 21 days and sells it on the 22nd day, booking a PNL due to price change and transaction cost. The actor-critic strategy rebalances the portfolio each day, incurring daily PNL and transaction costs. We also impose a **no-shorting** constraint that ensures position size does not become negative.

Details regarding the construction of the actor-critic strategy, training, testing, and comparison against the buy-and-hold strategy are provided below.

1. Daily log returns are computed using closing prices for each stock as shown in equation 5.85:

$$r_t = \log\frac{P_t}{P_{t-1}} \tag{5.85}$$

2. State of the actor-critic strategy is comprised of seven components:

   – **Variance ratio** of daily log returns. This is calculated by computing lagged variance of log returns over the last 21 days for each stock using equation 5.86:

$$\mu_t \quad = \frac{\sum_{k=t-21}^{t} r_k}{21}$$

$$\sigma_t^2 \quad = \frac{\sum_{k=t-21}^{t} \left(r_k - \mu_t\right)^2}{21} \tag{5.86}$$

Variance ratio is computed by dividing the variance calculated for a day by the variance calculated on the previous day as shown in equation 5.87:

$$VR_t = \frac{\sigma_t^2}{\sigma_{t-1}^2} \tag{5.87}$$

This has two values for each day, corresponding to each stock in the portfolio.

- **Correlation coefficient** $\rho_t$ of log returns for the stocks in the portfolio. Because we have two stocks, this is a scalar field calculated using equation 5.88, where $r_k$ and $\tilde{r}_k$ refer to log returns for the two stocks on day $k$:

$$\rho_t = \frac{\sum_{k=t-21}^{t} \left(r_k - \mu_t\right)\left(\tilde{r}_k - \tilde{\mu}_t\right)}{21\sqrt{\sigma_t^2 \tilde{\sigma}_t^2}} \tag{5.88}$$

- Two-day log return for each stock calculated using equation 5.89:

$$R_t = \log \frac{P_t}{P_{t-2}} \tag{5.89}$$

This quantity has two components, one for
each stock.

- **Relative weights** of each of the two stocks. The
weights sum to 1. Due to the no-shorting constraint, the weights must be positive.

3. Action is defined as the new portfolio weights.
To enforce the no-shorting constraint and the
requirement that weights add to 1, define an actor
network as a **softmax** layer with two outputs as the
final layer.

4. Reward is defined as shown in equation 5.90 for
a non-terminal time step. This consists of daily
PNL corresponding to portfolio rebalancing and
transaction costs. Terminal reward on the 21st day
is the value of the portfolio computed using closing
prices on that day and transaction costs involved
in liquidating the portfolio, as shown in equation
5.91. As before, $P_t$ and $\tilde{P}_t$ refer to the closing prices
of the two stocks. $\Delta N_1$ and $\Delta N_2$ refer to the shares
transacted for the two stock positions at time $t$:

$$W_t = \left(P_t - P_{t-1}\right)\Delta N_1 + \left(\tilde{P}_t - \tilde{P}_{t-1}\right)\Delta N_2 - \delta_{tr}\left(P_t \Delta N_1 + \tilde{P}_t \Delta N_2\right) \qquad (5.90)$$

$$W_T = \left(P_t - P_{t-1}\right)\Delta N_1 + \left(\tilde{P}_t - \tilde{P}_{t-1}\right)\Delta N_2 - \delta_{tr}\left(N_1 P_t + N_2 \tilde{P}_t\right) \qquad (5.91)$$

5. **Transaction costs** are assumed to be 0.1% of the
traded amount.

6. 75% of data is used for training and remaining 25%
for testing. Training data runs from year 2000 to the
beginning of 2016, and testing data runs from 2016
to 2022.

7. During training, each day the actor-critic strategy interacts with the environment, rebalancing its holdings as prescribed by the actor and earning a reward.

8. Discount factor $\gamma$ is 1.

9. During testing, the two strategies begin with the same portfolio value. After 21 days, their PNL is recorded. The two strategies are then traded beginning from the 22nd day to simulate real-world conditions.

10. Strategy PNL is added. Profits are not reinvested – a strategy that can boost returns for winners.

11. Total PNL for the two strategies is plotted in Figure 5-10. As can be seen, the actor-critic strategy's final PNL of $3.72 is higher than that of the buy-and-hold strategy's final PNL of $3.35.

12. The actor-critic strategy has a higher annualized Sharpe ratio of 1.4339 as compared with 1.0115 for the buy-and-hold strategy. The Sharpe ratio measures risk-adjusted average returns.

13. The actor-critic strategy outperforms buy-and-hold during COVID downturn, incurring a smaller loss as seen from Figure 5-10.

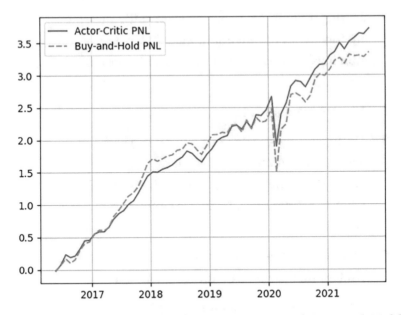

**Figure 5-10.** *Cumulative PNL for Actor-Critic and Buy-and-Hold Strategies on the Testing Dataset*

The code for this example is shown in Listing 5-14. Actual outputs for the actor-critic strategy may very slightly from those shown due to random weight initializers for network parameters.

**Listing 5-14.** Actor-Critic Trading Strategy Against the Buy-and-Hold Strategy

```
1   import os
2   from typing import List, Tuple
3
4   import matplotlib.pyplot as plt
5   import numpy as np
6   import pandas as pd
7   import seaborn as sns
8   import tensorflow as tf
9
```

```
10   from src.learner.ActorCriticLearner import
     AdvantageActorCriticLearner
11   from src.lib.ActorCriticNetwork import ACNetwork
12   from src.lib.Emulator import StateAndRewardEmulator
13   from src.lib.Episode import Episode
14   from src.lib.Sample import Sample
15
16
17   class PortfolioEmulator(StateAndRewardEmulator):
18       def __init__(self, dfs, var, covar, trx_cost, price_
         col, return_col, max_days, nstocks):
19           self.dfs = dfs
20           self.nStock = nstocks
21           self.var = var
22           self.covar = covar
23           self.iVar = 0
24           self.iCvar = nstocks
25           self.iRet = nstocks + nstocks*(nstocks - 1) // 2
26           self.iWeight = self.iRet + nstocks
27           self.priceCol = price_col
28           self.retCol = return_col
29           self.trxCost = trx_cost
30           self.maxDays = max_days
31           self._begin = 0
32           self._index = 0
33           self._state = None
34
35       def step(self, state, action):
36           pass
37
38       def setInitialState(self, state):
39           self._state = state
```

```
40
41          def setBeginIndex(self, value):
42              self._begin = value
43              self._index = value
44
45          def getReward(self, state, action, index, begin):
46              weights = state[self.iWeight:]
47              begin_price = np.array([df.loc[begin, self.
                    priceCol] for df in self.dfs])
48              price = np.array([df.loc[index, self.priceCol]
                    for df in self.dfs])
49              if index - begin == self.maxDays:
50                  nshares = np.divide(weights, begin_price) *
                        self.nStock
51                  pnl = (1 - self.trxCost) * np.sum(nshares * price)
52              else:
53                  new_position = action
54                  pos_change = new_position - weights
55                  nshares = np.divide(pos_change, begin_price)
                        * self.nStock
56                  price_change = np.array(
57                      [df.loc[index, self.priceCol] -
                          df.loc[index - 1, self.priceCol] for df
                          in self.dfs])
58                  pnl = np.sum(price_change * nshares) - self.
                        trxCost * np.sum(nshares * price)
59
60              return pnl
61
62          def tfEnvStep(self, action: tf.Tensor) -> List[tf.
                Tensor]:
63              self._index += 1
```

```
64          index = self._index
65          action = tf.squeeze(action)
66          weights = self._state[self.iWeight:]
67          price = [df.loc[index, self.priceCol] for df in
            self.dfs]
68          done = False
69          if index - self._begin == self.maxDays:
70              pnl = (1 - self.trxCost) * tf.reduce_
                sum(weights * price)
71              self._begin += 1
72              self._index = self._begin
73              done = True
74          else:
75              new_position = action
76              pos_change = new_position - weights
77              price_change = [df.loc[index, self.
                priceCol] - df.loc[index - 1, self.priceCol]
                for df in self.dfs]
78              pnl = tf.reduce_sum(price_change * pos_
                change) - self.trxCost * tf.reduce_sum(pos_
                change * price)
79
80          new_cvar = self.covar[index, :]
81          new_var = self.var[index, :]
82          new_ret = [df.loc[index, self.retCol] for df in
            self.dfs]
83          next_state = tf.concat((new_var, new_cvar, new_
            ret, action), axis=0)
84          self._state = next_state
85          return [next_state, pnl, done]
86
87
```

```
88   class PortOptim(object):
89       def __init__(self, stocks, inputdir, transaction_
         cost, training_data=0.75):
90           self.stocks = stocks
91           self.transactionCost = transaction_cost
92           self.nStock = len(stocks)
93           self.holdingPeriod = 21
94           self.dfs = []
95           self.priceCol = "Adj Close"
96           self.dateCol = "Date"
97           self.returnCol = "daily_return"
98
99           for stock in stocks:
100              filename = os.path.join(inputdir, "%s.csv"
                 % stock)
101              df = pd.read_csv(filename, parse_dates=[self.
                 dateCol])
102              df = self.calculateReturns(df)
103              self.dfs.append(df)
104
105          dates = self.dfs[0].loc[:, self.dateCol]
106          self.nDate = dates.shape[0]
107          for i in range(1, self.nStock):
108              self.dfs[i] = pd.merge(dates, self.dfs[i],
                 on=[self.dateCol], how="left")
109          self.nTrain = int(training_data * self.dfs[0].
             shape[0])
110          self.var, self.covar = self.calculateCovar()
111          self.emulator = PortfolioEmulator(self.dfs, self.
             var, self.covar, self.transactionCost,
```

```
112                                              self.priceCol,
                                                 self.returnCol,
                                                 self.
                                                 holdingPeriod,
                                                 self.nStock)
113        self.acnet = self.createActorCritic()
114        self.aclearner = AdvantageActorCriticLearner(sel
           f.acnet, discrete_actions=False)
115
116    def calculateReturns(self, df: pd.DataFrame) ->
       pd.DataFrame:
117        # 2 day return
118        price = df.loc[:, self.priceCol].values
119        df.loc[:, self.returnCol] = 0.0
120        logPrice = np.log(price)
121        logPriceDiff = logPrice[2:] - logPrice[0:-2]
122        df.loc[3:, self.returnCol] = logPriceDiff[0:-1]
123        return df
124
125    def calculateCovar(self) -> Tuple[np.ndarray,
       np.ndarray]:
126        dfs = self.dfs
127        variances = np.zeros((self.nDate, self.nStock),
           dtype=np.float32)
128
129        for index, df in enumerate(dfs):
130            ret = df.loc[:, self.returnCol].values
131            var = variances[:, index]
132            sum_val = np.sum(ret[2:2+self.holdingPeriod])
133            sumsq_val = np.sum(ret[2:2+self.
               holdingPeriod] * ret[2:2+self.holdingPeriod])
```

```
134              mean_val = sum_val / self.holdingPeriod
135              var[2+self.holdingPeriod-1] = sumsq_val /
                 self.holdingPeriod - mean_val * mean_val
136              for i in range(2+self.holdingPeriod, var.
                 shape[0]):
137                  sum_val += ret[i] - ret[i-self.
                     holdingPeriod]
138                  sumsq_val += ret[i] * ret[i] - ret[i-
                     self.holdingPeriod] * ret[i-self.
                     holdingPeriod]
139                  mean_val = sum_val / self.holdingPeriod
140                  var[i] = sumsq_val / self.holdingPeriod -
                     mean_val * mean_val
141
142          ncvar = self.nStock * (self.nStock - 1) // 2
143          covar = np.zeros((self.nDate, ncvar), dtype=np.
             float32)
144          count = 0
145
146          for i1 in range(self.nStock):
147              ret1 = self.dfs[i1].loc[:, self.
                 returnCol].values
148              for j in range(i1+1, self.nStock):
149                  ret2 = self.dfs[j].loc[:, self.
                     returnCol].values
150                  cvar = covar[:, count]
151                  for i in range(2 + self.holdingPeriod,
                     cvar.shape[0]):
152                      begin = i - self.holdingPeriod
153                      sum_val1 = np.sum(ret1[begin:begin +
                         self.holdingPeriod])
```

```
154            sum_val2 = np.sum(ret2[begin:begin +
               self.holdingPeriod])
155            mean_val1 = sum_val1 / self.
               holdingPeriod
156            mean_val2 = sum_val2 / self.
               holdingPeriod
157            sumprod = np.sum((ret1[begin:i] -
               mean_val1) * (ret2[begin:i] -
               mean_val2))
158            cvar[i] = sumprod / (self.
               holdingPeriod * variances[i, i1] *
               variances[i, j])
159
160                count += 1
161
162        # calculate variance ratio
163        variances[2+self.holdingPeriod+1:-1, :] =
           np.divide(variances[2+self.
           holdingPeriod+1:-1, :],
164                            variances[2+self.
                              holdingPeriod:-2, :])
165
166        return variances, covar
167
168    def createActorCritic(self):
169        value_network = tf.keras.models.Sequential()
170        # state: variance, cvar, ret, stock weights
171        ninp = self.nStock + self.nStock*(self.
           nStock-1)//2 + self.nStock + self.nStock
172        value_network.add(tf.keras.layers.Dense(4, input_
           shape=(ninp,)))
```

```
173         value_network.add(tf.keras.layers.Dense(10,
            activation="relu"))
174         value_network.add(tf.keras.layers.Dense(1))
175
176         anet = tf.keras.models.Sequential()
177         anet.add(tf.keras.layers.Dense(4, input_
            shape=(ninp,)))
178         anet.add(tf.keras.layers.Dense(10,
            activation="relu"))
179         anet.add(tf.keras.layers.Dense(self.nStock,
            activation="sigmoid"))
180         anet.add(tf.keras.layers.Softmax())
181
182         actor_optim = tf.keras.optimizers.Adam()
183         critic_optim = tf.keras.optimizers.Adam()
184
185         return ACNetwork(anet, value_network, self.
            emulator, 1.0, self.holdingPeriod, actor_optim,
            critic_optim)
186
187     def randomAction(self):
188         wts = np.random.random(self.nStock)
189         return np.divide(wts, wts.sum())
190
191     def getInitialWeights(self, day):
192         wts = [1.0/df.loc[day, self.priceCol] for df in
            self.dfs]
193         return np.divide(wts, np.sum(wts))
194
195     def generateTrainingEpisodes(self):
196         episodes = []
```

```
197            samples = [None]
198            begin = 0
199
200            for i in range(2 + self.holdingPeriod, self.
               nTrain - self.holdingPeriod):
201                curr_weights = self.getInitialWeights(begin)
202                rets = [df.loc[i, self.returnCol] for df in
                   self.dfs]
203                state = np.concatenate((self.var[i, :], self.
                   covar[i, :], rets, curr_weights))
204                action = self.randomAction()
205                reward = self.emulator.getReward(state,
                   action, i, begin)
206                if i - begin == self.holdingPeriod:
207                    begin = i
208                samples[0] = Sample(state, action,
                   reward, None)
209                episode = Episode(samples)
210                episodes.append(episode)
211
212            return episodes
213
214        def train(self):
215            # create episodes for training
216            episodes = self.generateTrainingEpisodes()
217            self.emulator.setBeginIndex(2+self.holdingPeriod)
218            self.aclearner.fit(episodes)
219
220        def actorCriticPnl(self, day):
221            pnl = -self.nStock
```

```
222            wts = np.full(self.nStock, 1.0/self.nStock,
               dtype=np.float32)
223            for i in range(day+1, day+self.holdingPeriod+1):
224                new_cvar = self.covar[i-1, :]
225                new_var = self.var[i-1, :]
226                new_ret = [df.loc[i-1, self.returnCol] for df
                   in self.dfs]
227                state = np.concatenate((new_var, new_cvar,
                   new_ret, wts))
228                vals = self.aclearner.predict(state)
229                abs_change = np.sum(np.abs(wts - vals[0]))
230                if abs_change > 0.1:
231                    wts[:] = vals[0]
232                pnl += self.emulator.getReward(state,
                   wts, i, day)
233        return pnl
234
235    def buyAndHoldPnl(self, day):
236        nstocks = [1.0 / df.loc[day, self.priceCol] for
               df in self.dfs]
237        price = [df.loc[day+self.holdingPeriod, self.
               priceCol] for df in self.dfs]
238        return -self.nStock + (1 - self.transactionCost)
               * np.sum(np.multiply(price, nstocks))
239
240    def test(self):
241        pnl_data = []
242        pnl_bh_data = []
243        pnl_diff = []
244        dates = []
245        self.emulator.setBeginIndex(self.nTrain)
```

```
246        for i in range(self.nTrain, self.nDate-self.
           holdingPeriod-2, self.holdingPeriod):
247            pnl = self.actorCriticPnl(i)
248            pnl_bh = self.buyAndHoldPnl(i)
249            pnl_diff.append(pnl - pnl_bh)
250            pnl_data.append(pnl)
251            pnl_bh_data.append(pnl_bh)
252            dates.append(self.dfs[0].loc[i, self.
               dateCol])
253
254

255        perf_df = pd.DataFrame(data={"Actor-Critic PNL":
           np.cumsum(pnl_data),
256                                "Buy-and-Hold PNL":
                                   np.cumsum(pnl_
                                   bh_data)},
257                            index=np.array(dates))
258        final_pnl = np.array([np.sum(pnl_data),
           np.sum(pnl_bh_data)])
259        mean_pnl = np.array([np.mean(pnl_data),
           np.mean(pnl_bh_data)])
260        sd_pnl = np.array([np.std(pnl_data), np.std(pnl_
           bh_data)])
261        sr = np.sqrt(252.0/self.holdingPeriod) * mean_
           pnl/sd_pnl
262        print("Actor-Critic: final PNL: %f, SR: %f" %
           (final_pnl[0], sr[0]))
263        print("Buy-and-hold: final PNL: %f, SR: %f" %
           (final_pnl[1], sr[1]))
264        sns.lineplot(data=perf_df)
265        plt.grid(True)
```

```
266              plt.show()
267
268
269    if __name__ == "__main__":
270          stocks = ["MSFT", "BA"]
271          inputdir = r"C:\prog\cygwin\home\samit_000\RLPy\
             data\stocks"
272          portopt = PortOptim(stocks, inputdir, 0.001)
273          portopt.train()
274          portopt.test()
```

## 5.5.3 Natural Actor-Critic Algorithms

Natural actor-critic algorithms address the problem of slow convergence witnessed by many actor-critic algorithms that are based on stochastic gradient descent. Gradient descent has a linear convergence rate. This can be improved to quadratic convergence by using a Newton step as shown in equation 5.92 that uses the Hessian $\dfrac{\partial^2 f}{\partial w^2}$. Natural actor-critic algorithms are motivated by using a Newton step for updating the parameters instead of relying upon gradient descent.

$$\min_{w} f(w)$$

$$w_{new} \leftarrow w_{old} - \left(\frac{\partial^2 f}{\partial w^2}\right)^{-1} \frac{\partial f}{\partial w} \tag{5.92}$$

More formally, natural gradient can be derived by minimizing the sum of square deviations of the actual advantage function $Q^\pi(s, a) - V^\pi(s)$ from that predicted by the parameterized advantage function $w'\psi_{sa}$ for all states. The proof is illustrated in equation 5.94. We have also used the fact that the advantage function is compatible as defined in equation 5.76. Using the parameterization for the advantage function, we can write equation 5.93, which has been used in the derivation shown in equation 5.94:

$$\frac{\partial f(s,a,\mathbf{w})}{\partial \mathbf{w}} = \psi_{sa} = \frac{\partial \pi(s,a,\theta)}{\partial \theta} \frac{1}{\pi(s,a,\theta)} \tag{5.93}$$

$$\min_{\mathbf{w}} E^{\pi}(w) = \sum_{s\in S} d^{\pi}(s) \sum_{a\in A} \pi(s,a)[Q^{\pi}(s,a) - V^{\pi}(s) - w'\psi_{sa}]^2$$

$$\Rightarrow -2\sum_{s\in S} d^{\pi}(s) \sum_{a\in A} \pi(s,a)[Q^{\pi}(s,a) - V^{\pi}(s) - w'\psi_{sa}]\psi_{sa} = 0$$

$$\Rightarrow \sum_{s\in S} d^{\pi}(s) \sum_{a\in A} \pi(s,a)Q^{\pi}(s,a) - \sum_{s\in S} d^{\pi}(s)V^{\pi}(s)$$

$$\sum_{a\in A} \pi(s,a)\psi_{sa} = \sum_{s\in S} d^{\pi}(s) \sum_{a\in A} \pi(s,a)w'\psi'_{sa}\psi_{sa}$$

$$\Rightarrow \sum_{s\in S} d^{\pi}(s) \sum_{a\in A} \pi(s,a)Q^{\pi}(s,a) - \sum_{s\in S} d^{\pi}(s)V^{\pi}(s)\sum_{a\in A} \pi(s,a)$$

$$\frac{\partial \pi(s,a)}{\partial \theta} \frac{1}{\pi(s,a)} = w' \sum_{s\in S} d^{\pi}(s) \sum_{a\in A} \pi(s,a)\nabla'_{\theta}\psi_{sa}\nabla_{\theta}\psi_{sa}$$

$$\Rightarrow E_{s\sim d^{\pi},a\sim\pi}[Q^{\pi}(s,a)] - \sum_{s\in S} d^{\pi}(s)V^{\pi}(s)\frac{\partial \sum_{a\in A} \pi(s,a)}{\partial \theta} =$$

$$w' \sum_{s\in S} d^{\pi}(s) \sum_{a\in A} \pi(s,a)\psi_{sa}\psi_{sa}$$

$$\Rightarrow w = [\sum_{s\in S} d^{\pi}(s) \sum_{a\in A} \pi(s,a)\psi_{sa}\psi_{sa}]^{-1} E_{s\sim d^{\pi},a\sim\pi}[Q^{\pi}(s,a)]$$

$$\Rightarrow w = G^{-1}(\theta)E_{s\sim d^{\pi},a\sim\pi}[Q^{\pi}(s,a)] \tag{5.94}$$

$G(\theta)$ in equation 5.94 is the Fisher information matrix.

## 5.5.4 Cross Entropy–Based Actor-Critic Algorithms

A cross entropy–based actor-critic algorithm uses the cross entropy method illustrated in the section "Policy Learning Using Cross Entropy" to train a deterministic policy, called $\pi_{CEM}$. To train $\pi_{CEM}$, it uses the existing state-action value function $Q(s, a, \theta_c)$ in cross entropy optimization. The actor uses the cross entropy policy $\pi_{CEM}$ to train the deterministic policy

$\pi(s,\theta)$ using supervised learning using equation 5.95. The critic uses $\pi_{CEM}$ to update the parameters $\theta_c$ of the state-action value function $Q(s,a,\theta_c)$ using equation 5.96:

$$\min_{\theta_a}\left[\pi_{CEM}(s)-\pi(s,\theta_a)\right]^2$$

$$\theta_a^{new}=\theta_a^{old}+\alpha_a\left[\pi_{CEM}(s)-\pi(s,\theta_a)\right]\frac{\partial\pi(s,\theta_a)}{\partial\theta_a} \tag{5.95}$$

$$\min_{\theta_c}\left[r(s',s,a)+\gamma Q(s',\pi_{CEM}(s'),\theta_c^-)-Q(s,a,\theta_c)\right]^2$$

$$\theta_c^{new}=\theta_c^{old}+\alpha_c\left[r(s',s,a)+\gamma Q(s',\pi_{CEM}(s'),\theta_c^-)-Q(s,a,\theta_c)\right]\frac{\partial Q(s,a,\theta_c)}{\partial\theta_c} \tag{5.96}$$

Instead of relying on an $\epsilon$-greedy policy to construct a target value for the critic, the cross entropy–guided critic uses $\pi_{CEM}$ – the policy derived using the cross entropy method. To ensure stability of the learning process, one has to use a replay buffer and a target network in the critic, as was described in the section "Double Q-Learning."

# CHAPTER 6

# Recent RL Algorithms

Reinforcement learning has taken some of its biggest strides in the past few years, with innovative algorithms paving the way for reinforcement learning to beat human opponents at many games and creating new performance benchmarks. This chapter describes the recent advances in reinforcement learning algorithm development, explaining the underlying theory and elucidating it with additional examples. In doing so, it provides valuable insights into the theory and implementation of these algorithms and a window into using these algorithms in practical problems.

## 6.1 Double Deep Q-Network: DDQN

DDQN was proposed in 2016 by H. Hasselt, A. Guez, and D. Silver in a paper titled "Deep Reinforcement Learning with Double Q-Learning." Q-learning suffers from overestimation of correction in stochastic gradient descent. This overestimation occurs because the same Q-network is employed for estimating the optimum action using a greedy approach and evaluating the value function. If one action value becomes larger than other action values for a specific state, a greedy policy will pick that action, and parameter updates will further increase the state-action value. The DDQN algorithm handles the problem of overestimation by using double Q-learning. This algorithm generalizes double Q-learning by using deep Q-networks.

© Samit Ahlawat 2023
S. Ahlawat, *Reinforcement Learning for Finance*,
https://doi.org/10.1007/978-1-4842-8835-1_6

A deep Q-network represents the action value function as $Q(s, a, \theta)$, where $\theta$ is a set of parameters to be learned by the algorithm. The Q-learning algorithm updates the parameter values using a TD(0) (or TD(n)) target, as shown in equation 6.1:

$$\theta \leftarrow \theta + \alpha \left( Q_{target} - Q(s_t, a_t, \theta) \right) \nabla_\theta Q(s_t, a_t, \theta)$$

$$Q_{target} = r(s_{t+1}, s_t, a_t) + \gamma Q\left( s_{t+1}, \underset{a}{\mathrm{argmax}}\, Q(s_{t+1}, a, \theta), \theta \right) \tag{6.1}$$

If there is an overestimation in $Q(s_{t+1}, a, \theta)$ making its action value bigger than that of other action values in that state, $\mathrm{argmax}_a Q(s_{t+1}, a, \theta)$ will select that action, overestimating $\theta$. Overestimated $\theta$ will drive further overestimation in action values in subsequent iterations.

Let us numerically establish the occurrence of overestimation in Q-learning. Let us consider a state $s$ with equal action values for all actions. This implies that the state value function $V(s)$ is equal to the state-action value function $Q(s, a)$ for that state, as shown in equation 6.2. We denote the actual (true) action value function as $Q(s, a)$ and estimated action value function as $\hat{Q}(s,a)$:

$$Q(s,a) = q \text{ for some } s \text{ and all } a \in \mathcal{A}$$
$$\Rightarrow V(s) = \Sigma_{a \in A} \pi(s,a) Q(s,a) \tag{6.2}$$
$$= q \Sigma_{a \in A} \pi(s,a) = q = Q(s,a)$$

Let us assume that estimation errors $\epsilon_a(s)$ are unbiased (i.e., have a zero mean) and are independent and identically distributed as a uniform random number in the range $[-1, 1]$, as shown in equation 6.3:

$$\epsilon_a(s) = \hat{Q}(s,a) - Q(s,a) = \hat{Q}(s,a) - q$$
$$\epsilon_a(s) \sim \mathcal{U}[-1,1] \tag{6.3}$$

Even when individual errors $\epsilon_a(s)$ are small, $E\left[ \max_a \hat{Q}(s,a) - V(s) \right]$

can be large and close to 1. This can be seen from equation 6.4. $\frac{m-1}{m+1}$ can be close to 1 for large action spaces with large m, even when individual errors are small. This explains why Q-learning is prone to overestimation errors:

$$\max_a \hat{Q}(s,a) - V(s) = \max_a \varepsilon_a(s)$$

$$m = \text{number of actions}$$

$$\varepsilon_a(s) \sim \mathcal{U}[-1,1]$$

$$E\left[\max_a \hat{Q}(s,a) - V(s)\right] = E\left[\max_a \varepsilon_a(s)\right]$$

$$P\left[\varepsilon_{ai}(s) \leq x\right] = \frac{1+x}{2} \text{ for uniformly distributed } \varepsilon_{ai}(s) \tag{6.4}$$

$$\Rightarrow P\left[\max_a \varepsilon_a(s) \leq x\right] = P\left[\varepsilon_{aj}(s) = x\right]\prod_{k \neq j}P\left[\varepsilon_{ak}(s) \leq x\right]$$

$$= dx\left(\frac{1+x}{2}\right)^{m-1}$$

$$\Rightarrow E\left[\max_a \varepsilon_a(s)\right] = \int_{-1}^{1}x\left(\frac{1+x}{2}\right)^{m-1}dx = \frac{m-1}{m+1}$$

Estimation error is directly related to the correction applied by Q-learning, as shown in equation 6.5:

$$\Delta\theta = \alpha\left[r + \gamma\max_a \hat{Q}(s_{t+1},a,\theta) - V(s_t)\right]$$

$$= \alpha\left[r + \gamma\max_a \hat{Q}(s_{t+1},a,\theta) - \left(r + \gamma V(s_{t+1})\right)\right]$$

$$= \alpha\gamma\left[\max_a \hat{Q}(s_{t+1},a,\theta) - V(s_{t+1})\right] \tag{6.5}$$

$$= \alpha\gamma\max_a \varepsilon_a(s_{t+1})$$

Equation 6.4 shows that expected values of correction used in Q-learning can be much larger than individual errors, and equation 6.5 shows that estimation errors result in overestimation of parameters. We can also show that individual errors calculated by Q-learning are susceptible to over-correction if some action value is overestimated.

Double Q-learning solves the problem of overestimation by using two networks: a learned network that is used for action evaluation and update of weights and a target network that is updated only periodically and is used for constructing the target value used in gradient descent. Let us denote the parameters of the learned network by $\theta$ and the parameters of target network by $\theta^-$, with the negative superscript indicating that the target network parameters are updated with a lag. The update rule for target and learned network parameters followed by double Q-learning is illustrated in equation 6.6:

$$\theta \leftarrow \theta + \alpha \left[ \left( r + \gamma Q(s_{t+1}, \arg\max_a Q(s_{t+1}, a, \theta), \theta^-) \right) - Q(s_t, a_t, \theta) \right] \nabla_\theta Q(s_t, a_t, \theta)$$

$$\theta^- \leftarrow \theta \text{ with a delay, e.g. if mod } T_{period} = 0 \tag{6.6}$$

Instead of using a periodic update of target network parameters from learned network parameters as shown in equation 6.6, one can alternatively use an update rule shown in equation 6.7 where the target parameters are updated periodically from learned parameters with a weight. $\beta = 0.9$ is commonly used in this update:

$$\theta^- \leftarrow \beta\theta^- + (1-\beta)\Delta\theta$$

$$\beta \approx 1 \tag{6.7}$$

$$\Delta\theta = \alpha \left[ \left( r + \gamma Q(s_{t+1}, \arg\max_a Q(s_{t+1}, a, \theta), \theta^-) \right) - Q(s_t, a_t, \theta) \right] \nabla_\theta Q(s_t, a_t, \theta)$$

Since the target network is updated periodically, any overestimates coming from parameter update do not immediately affect the target value function. To establish this numerically, let us examine the error for the case where all action values in a state are equal, as shown in equation 6.8:

$$\epsilon_a = Q\left( s_t, \arg\max_a Q(s_t, a, \theta), \theta^- \right) - V(s)$$

$$= Q(s_t, a, \theta^-) - V(s) \tag{6.8}$$

Since $Q(s, a, \theta^-)$ is the action value network with an old value of $\theta$, $\epsilon_a$ is a random number from a uniform distribution $[-1, 1]$ according to the assumed distribution for errors. Calculating the expected value of this error, we see that it has a mean of zero and a standard deviation of $1/3$ as shown in equation 6.9. This shows that Q-learning is able to avoid overestimates from propagating. While the original error $\epsilon_a$ was uniformly distributed with mean 0 and standard deviation of $1/\sqrt{3}$, the expected error propagated when using the double Q-network has mean 0 and a standard deviation of $1/3$:

$$\epsilon_a = Q\left(s_t, \operatorname*{argmax}_a Q(s_t, a, \theta), \theta^-\right) - V(s) \sim Q\left(s_{t,\prime\prime}, \theta^-\right) - V(s)$$

$$E\left[Q\left(s_{t,\prime\prime}, \theta^-\right) - V(s)\right] = \int_{-1}^{1} x\left(\frac{1+x}{2}\right) dx = 0 \tag{6.9}$$

$$E\left[Q\left(s_{t,\prime\prime}, \theta^-\right) - V(s)\right]^2 = \int_{-1}^{1} x\left(\frac{1+x}{2}\right)^2 dx = \frac{1}{3}$$

In order to ensure that the mini-batch of samples used to perform gradient descent are uncorrelated, DDQN uses a replay buffer of experiences. The DDQN algorithm is shown in pseudo-code 13.

# 6.2 Balancing a Cartpole Using DDQN

Let us apply double deep Q-network to solve the cartpole balancing problem described in the last chapter. The only code change required is to instantiate and use a DDQN agent instead of a DQN agent as shown in Listing 6-1.

*Listing 6-1.* Balancing a Cartpole Using DDQN

```
1    import numpy as np
2    import tensorflow as tf
3    from src.learner.DDQN import DDQN
4
```

```
5    tf.random.set_seed(10)
6    np.random.seed(10)
7
8
9    class CartpoleVoDDQN(CarpoleVoDQN):
10       def createAgent(self):
11           replay_buf = MostRecentReplayBuffer(2 * self.
             minibatchSize)
12           return DDQN(self.qfunc, self.emulator, self.
             nfeatures, self.nactions,
13                       replay_buffer=replay_buf, discount_
             factor=self.discountFactor,
14                       minibatch_size=20, epochs_training=20,
             sync_period=2)
15
16
17   if __name__ == "__main__":
18       cartpole = CartpoleVoDDQN()
19       cartpole.balance()
```

---

**Algorithm 13**    Double Deep Q-Network Learning

**Require**: Action value function parameterized by $\theta$: $Q(s, a, \theta)$, replay buffer $\mathcal{B}$, batch size $N_b$, update frequency $N_t$, discount factor $\gamma$, and learning rate $\alpha$

1: Create a target network $Q(s, a, \theta^-)$ by copying the parameters $\theta^- \leftarrow \theta$.

2: **for** each episode in episodes **do**

3:    Initialize the starting state $s_0$ using the episode's starting state.

4:    count $\leftarrow 0$

5:    **for** each t = 0; 1; 2; $\cdots$ in the episode **do**

6:        count $\leftarrow$ count $+1$

7:        Select action $a_t$ using an $\epsilon$-greedy policy: with probability $\epsilon$, $a_t$ is a random action sampled from $\mathcal{A}$ with probability $1 - \epsilon$, $a_t = \text{argmax}_a Q(s_t, a, \theta)$.

8:        Take action $a_t$, observe $r_t$, and transition to state $s_{t+1}$.

9:        Add $(s_t, a_t, r_t, s_{t+1})$ to replay buffer $\mathcal{B}$.

10:        Sample a random mini-batch of size $N_b$ from replay buffer $\mathcal{B}$.

11:        *descent* $\leftarrow 0$

12:        **for** each $(s, a, r, s')$ in the mini-batch **do**

13:            Calculate the action
$a^* = \text{arg max}_a Q(s, a, \theta). a^* = \text{argmax}_a Q(s, a, \theta)$

14:            Take action $a^*$, observe reward $r^*$, and transition to state $s^*$.

15:            Add $(s, a^*, r^*, s^*)$ to replay buffer $\mathcal{B}$.

16:            Calculate target $y = r + \gamma Q(s^*, a^*, \theta^-)$ if $s$ is non-terminal or $y = r$ if $s$ is terminal.

17:            descent $\leftarrow$ descent $+(y - Q(s, a, \theta))\nabla_\theta Q(s, a, \theta)$

18:        **end for**

19:        descent $\leftarrow \dfrac{\text{descent}}{N_b}$

20:        $\theta \leftarrow \theta + \alpha$ descent

21:                    $\theta^- \leftarrow \theta$ if count mod $N_t = 0$

22:    **end for**

23: **end for**

---

Results show that most cartpole balancing episodes reach the maximum length of 50 samples using the DDQN agent, outperforming the random agent.

# 6.3 Dueling Double Deep Q-Network

Dueling DDQN was introduced by Wang et al. (2016) in a paper titled "Dueling Network Architectures for Deep Reinforcement Learning." This algorithm uses a new network architecture for calculating the action value function $Q(s, a)$ by simultaneously calculating the state value function $V(s)$ and advantage function $A(s, t) = Q(s, t) - V(s)$. The action value function is computed by adding the state value function and advantage function.

For a deterministic optimum policy, the value function is equal to the action value function. This can be seen from equation 6.10:

$$
\begin{aligned}
Q^\pi(s,a) &= E\left[\Sigma_{t=0}^\infty \gamma^t r_t\right] = E_{s'}\left[r + \gamma E_{a' \sim \pi(s')}\left[Q^\pi(s',a')\right]\right] \\
&= E_{s'}\left[r + \gamma V^\pi(s')\right] \\
V^\pi(s) &= E_{a \sim \pi(s)}\left[Q^\pi(s,a)\right] \\
&\text{optimal } Q^*(s,a) = \max_\pi Q^\pi(s,a) \text{ with } a = \underset{a}{\operatorname{argmax}} Q^*(s,a) \\
&\Rightarrow V^*(s) = Q^*(s,a) \text{ for optimal, deterministic policy} \\
&\Rightarrow A^*(s,a) = 0
\end{aligned}
\tag{6.10}
$$

Dueling DDQN uses a network with a shared section for calculating the value function and advantage function and a section that is not shared between these two value functions. This can be written explicitly as shown in equation 6.11. Upon convergence to an optimum policy, advantage $A(s, a, \theta, \alpha)$ will be 0. $\theta$ denotes the common parameters between value function and advantage function networks. $\beta$ and $\alpha$ denote the non-shared parameters of value function and advantage function networks:

$$Q(s,a,\theta,\alpha,\beta) = V(s,\theta,\beta) + A(s,a,\theta,\alpha) \tag{6.11}$$

In order to enhance the stability of the learning process, the authors propose two alternatives A1 and A2 for calculating the advantage function as shown in equation 6.12:

$$\begin{aligned}
A1 &: Q(s,a,\theta,\alpha,\beta) = V(s,\theta,\beta) + \left( A(s,a,\theta,\alpha) - \max_{a'} A(s,a',\theta,\alpha) \right) \\
A2 &: Q(s,a,\theta,\alpha,\beta) = V(s,\theta,\beta) + \left( A(s,a,\theta,\alpha) - \frac{1}{size(\mathcal{A})} \Sigma_{a'} A(s,a',\theta,\alpha) \right)
\end{aligned} \tag{6.12}$$

# 6.4 Noisy Networks

Noisy networks for exploration were proposed in a paper titled "Noisy Networks for Exploration" in 2017 by M. Fortunato et al. The authors proposed adding noise to weights of a deep learning agent to facilitate exploration of state space. The parameters of the noise-generating process are learned along with network weights using gradient descent. Using this approach, they found that a reinforcement learning agent can surpass humans in playing Atari games.

Reinforcement learning presents an exploration vs. exploitation dilemma. Exploiting the information learned so far would lead the learner to take steps found to yield greatest rewards in past experiences. However, it could also lead to the learner getting stuck in local optima because it did not explore the entire state-action space. Exploring the state-action space requires the learner to visit the previously unexplored state-action space in order to get a more complete view of the rewards available, often taking actions that appear to be suboptimal because the estimate of rewards is incomplete due to insufficient exploration of the state-action space. Traditional approaches to dealing with this dilemma include using an $\epsilon$-greedy policy and entropy regularization. An $\epsilon$-greedy policy resorts to exploitation, but occasionally switches to exploration by selecting a random action with some specified probability. Entropy regularization adds an entropy term to the objective function of discounted future rewards in order to visit unexplored regions of the state-action space. Entropy $H_\pi$ of a policy is given by equation 6.13:

$$H_\pi(s_t) \ = -\sum_{a_t} \pi(s_t, a_t) \log(\pi(s_t, a_t))$$
$$= -E_{a \sim \pi}\left[\log(\pi(s_t, a_t))\right]$$

(6.13)

Entropy of a policy is maximum when probabilities of taking different actions in a state are equal, that is, when $\pi(s_t, a_t) = \pi(s_t)$ for all $a_t$. The discounted reward function under entropy regularization is modified to include the entropy term, as shown in equation 6.14, and is called entropy-augmented discounted rewards, $J_{ENT}^\pi$. $\tau$ is a weighing factor that determines the relative importance of the entropy regularization term. When a learner has explored only a few action-state combinations, probabilities of explored actions are non-zero, while the remaining actions have zero probability resulting in a low entropy. This decreases the entropy contribution to the discounted reward for that action-state combination, favoring exploration of other actions. Policy gradient for entropy-regularized discounted rewards is shown in equation 6.15:

$$J_{ENT}^\pi(s_0) = \sum_{t_0}^{\infty} \gamma^t \left[r(s_t, a_t) + \tau H_\pi(s_t)\right]$$

(6.14)

$$\frac{\partial J_{ENT}^\pi(s_0)}{\partial \theta} = \sum_{s \in S} d^\pi(s|s_0) \sum_{a \in A} \left[\frac{\partial \pi(a|s, \theta)}{\partial \theta} + \tau \frac{\partial H_\pi(s)}{\partial \theta}\right] A^\pi(s, a)$$

(6.15)

In many applications, an $\epsilon$-greedy policy and an entropy regularization approach result in small changes to parameters that do not result in efficient exploration of state-action space. Noisy networks address this shortcoming by adding a noise generated by a parameterized process to network weights and learning the network weights and noise parameters using gradient descent.

# 6.5 Deterministic Policy Gradient

In the formulation of the policy gradient theorem in equation 5.67, the policy is stochastic. Stochastic policy prescribes a probability of taking an action in a given state, $\pi(a|s,\theta)$, where $\theta$ is the set of policy parameters. For continuous action spaces, this means we must discretize the action space or work with integrals over the action space. Both of these alternatives have disadvantages. Discretizing a continuous action space makes the approach vulnerable to the curse of dimensionality for high-dimensional action spaces. In addition, one must discretize the action space at a fine enough level to be able to discriminate between different actions of interest. Integrating over a continuous action space is often intractable, necessitating the use of approximate numerical techniques. In practice, one often discretizes a continuous action space and uses random sampling with probability prescribed by the stochastic policy to select the next action. Many practical problems are modeled more effectively by using a continuous action space with deterministic policy.

In 2014, D. Silver et al. published a paper titled "Deterministic Policy Gradient Algorithms" in which they proposed an algorithm for learning a deterministic policy over continuous action spaces using policy gradient. For a stochastic policy, equation 6.16 shows the policy gradient theorem (following directly from equation 5.67). $J(\pi_\theta)$ is the performance function used in policy gradient:

$$\frac{\partial J(\pi_\theta)}{\partial \theta} = \sum_{s \in S} d^\pi(s|s_0) \sum_{a \in A} \frac{\partial \pi(a|s,\theta)}{\partial \theta} Q^\pi(s,a)$$

$$= E_{s \sim d^\pi, a \sim \pi_\theta} \left[ \nabla_\theta \log(\pi(a|s,\theta)) Q^\pi(s,a) \right] \qquad (6.16)$$

In order to formulate an equation for deterministic policy gradient, let us examine how an off-policy actor-critic algorithm updates network parameters in the next subsection.

## 6.5.1  Off-Policy Actor-Critic Algorithm

Policy gradient can be estimated off-policy in an actor-critic algorithm. We use a behavior policy, $\beta$, to generate samples over the state distribution and use it to evaluate the value function of the target policy. For continuous action space, the reward function used by the actor is given by equation 6.17:

$$J_\beta(\pi_\theta) = \iint_{SA} d^\beta(s) \pi_\theta(a|s) Q^\pi(s,a) \, da \, ds \qquad (6.17)$$

A derivative of the reward function is used to update the policy function parameters using a stochastic gradient ascent approach as shown in equation 6.18. This is the update performed by the actor. $Q^\pi(s, a)$ is not known and is replaced by the advantage function calculated using the value function $V_\phi(s)$ from the critic. Policy $\beta_\theta(a|s)$ is used to generate episodes, and we are trying to learn the target policy $\pi_\theta(a|s)$:

$$\nabla_\theta J_\beta(\pi_\theta) = \iint_{SA} \nabla_\theta \pi_\theta(a|s) Q^\pi(s,a) \, da \, ds$$

$$= E_{s \sim d^\beta, a \sim \beta} \left[ \frac{\pi_\theta(a|s)}{\beta_\theta(a|s)} \nabla_\theta \log(\pi_\theta(a|s))(r' + \gamma V_\phi(s') - V_\phi(s)) \right] \qquad (6.18)$$

$$\theta \leftarrow \theta_a + \alpha \nabla_\theta J_\beta(\pi_\theta)$$

The critic updates the parameters $\phi$ of the value function using supervised learning. The target can be constructed using TD(0) expansion, as shown in equation 6.19. Like in actor updates, we have to use the weight $\dfrac{\pi_\theta(a|s)}{\beta_\theta(a|s)}$ to account for the fact that we are sampling from behavior policy $\beta$, which is different from the target policy $\pi$:

$$\phi = \operatorname{argmin}_\phi \sum \left[ \frac{\pi_\theta(a|s)}{\beta_\theta(a|s)}\left(r+\gamma V_\phi(s')\right) - V_{\phi(s)} \right]^2$$

$$\delta = \left[ \frac{\pi_\theta(a|s)}{\beta_\theta(a|s)}\left(r+\gamma V_\phi(s')\right) - V_{\phi(s)} \right] \tag{6.19}$$

$$\phi \leftarrow \phi + \alpha_c \delta \nabla_\phi V_\phi(s)$$

## 6.5.2 Deterministic Policy Gradient Theorem

For deterministic policy $\pi_\theta$, the reward function $J(\pi_\theta)$ can be written as shown in equation 6.20. Because the policy is deterministic, there is no inner integral over the action space. Taking the derivative of the reward function with respect to $\theta$ gives the deterministic policy gradient:

$$J(\pi_\theta) = \int_S d^\pi(s) Q^\pi(s, \pi_\theta(s)) ds$$

$$\nabla_\theta J(\pi_\theta) = \int_S d^\pi(s) \nabla_\theta \pi_\theta(s) \nabla_\pi Q^\pi(s, \pi_\theta(s)) ds \tag{6.20}$$

$$= E_{s \sim d^\pi, a=\pi_\theta}\left[ \nabla_\theta \pi_\theta(s) \nabla_\pi Q^{\pi_\theta}(s, \pi_\theta(s)) \right]$$

Equation 6.20 can be used to derive gradient ascent–based parameter update for the actor to formulate an actor-critic algorithm based on deterministic policy.

# 6.6 Trust Region Policy Optimization: TRPO

Trust region policy optimization optimizes a policy using policy iteration, with the actual implementation relying on Monte Carlo samples from the policy. It ensures guaranteed monotonic improvement of the policy with each iteration. This algorithm was proposed by J. Schulman et al. in a paper titled "Trust Region Policy Optimization" in 2015.

Let $\pi$ represent a stochastic policy $\pi : S \times A \to [0,1]$. As before, $J(\pi)$ denotes the expected discounted reward. We can write the expected discounted reward of policy $\pi$ in terms of expected discounted reward of another stochastic policy $\tilde{\pi}$ as shown in equation 6.21. The proof of this proposition is shown in equation 6.22:

$$J(\pi) = E_{s_0 \sim \rho_0, a_t \sim \pi, s_t \sim P(s|s_{t-1}, a_t)} \left[ \gamma^t r(s_t) \right]$$

$$J(\tilde{\pi}) = J(\pi) + E_{a_t \sim \tilde{\pi}} \left[ \sum_{t=0}^{\infty} \gamma^t A^\pi (s_t, a_t) \right] \tag{6.21}$$

$$A^\pi (s_t, a_t) = E[r(s_t) + \gamma V^\pi (s_{t+1}) - V^\pi (s_t)]$$

$$\Rightarrow E_{a \sim \tilde{\pi}, s_0} \left[ \sum_{t=0}^{\infty} \gamma^t A^\pi (s_t, a_t) \right] = E_{a \sim \tilde{\pi}} \left[ \sum_{t=0}^{\infty} \gamma^t \left( r(s_t) + \gamma V^\pi (s_{t+1}) - V^\pi (s_t) \right) \right]$$

$$= E_{a \sim \tilde{\pi}, s_0} \left[ \sum_{t=0}^{\infty} \gamma^t r(s_t) + \sum_{t=1}^{\infty} \gamma^t V^\pi (s_{t+1}) - \sum_{t=0}^{\infty} \gamma^t V^\pi (s_t) \right] \tag{6.22}$$

$$= E_{a \sim \tilde{\pi}, s_0} \left[ \sum_{t=0}^{\infty} \gamma^t r(s_t) - V^\pi (s_0) \right]$$

$$= J(\tilde{\pi}) - J(\pi)$$

$$\Rightarrow J(\tilde{\pi}) = J(\pi) + E_{a \sim \tilde{\pi}} \left[ \sum_{t=0}^{\infty} \gamma^t A^\pi (s_t, a_t) \right]$$

Expected discounted reward of policy $\tilde{\pi}$ can be rewritten as a summation over states, as shown in equation 6.24. $\rho\tilde{\pi}$ denotes the discounted state visitation frequency as shown in equation 6.23. From equation 6.24, one can observe that a policy update $\pi \to \tilde{\pi}$ improves the expected discounted reward if the expected advantage $\sum_{a \in A} \tilde{\pi}(a|s) A^{\pi}(s,a)$ is non-negative at every state:

$$\rho^{\tilde{\pi}} = P(s_0) + \gamma P(s_1|a_0 \sim \tilde{\pi}) + \gamma^2 P(s_2|a_1 \sim \tilde{\pi}) + \cdots \tag{6.23}$$

$$J(\tilde{\pi}) = J(\pi) + \sum_{s \in S} \rho^{\tilde{\pi}}(s) \sum_{a \in A} \tilde{\pi}(a|s) A^{\pi}(s,a) \tag{6.24}$$

In order to render equation 6.24 amenable to use in updating initial policy $\pi$ to improved policy $\tilde{\pi}$ with higher expected discounted reward, it can be approximated as shown in equation 6.25 using discounted state visitation frequency $\rho^{\pi}$. However, this approximation is only valid for small changes to policy $\pi \to \tilde{\pi}$. Schulman et al. proved the bound on expected discounted reward of improved policy $J(\tilde{\pi})$ as shown in equation 6.26, which is the foundation of the trust region policy optimization algorithm. As before, $\tilde{\pi}$ denotes the updated policy. $D_{KL}(\pi,\tilde{\pi})$ denotes the Kullback-Leibler divergence between the old policy $\pi$ and the new policy $\tilde{\pi}$:

$$\begin{aligned} L^{\pi}(\tilde{\pi}) &= J(\pi) + \sum_{s \in S} \rho^{\tilde{\pi}}(s) \sum_{a \in A} \tilde{\pi}(a|s) A^{\pi}(s,a) \\ &\approx J(\pi) + \sum_{s \in S} \rho^{\pi}(s) \sum_{a \in A} \tilde{\pi}(a|s) A^{\pi}(s,a) \end{aligned} \tag{6.25}$$

$$\begin{aligned} J(\tilde{\pi}) &\geq L^{\pi}(\tilde{\pi}) - CD_{KL}^{\max}(\pi,\tilde{\pi}) \\ C &= \frac{4\epsilon\gamma}{(1-\gamma)^2} \\ \epsilon &= \max_{s,a} A^{\pi}(s,a) \\ D_{KL}^{\max}(\pi,\tilde{\pi}) &= \max_{s} D_{KL}(\pi(.|s) \| \tilde{\pi}(.|s)) \\ &= \max_{s} \sum_{a \in A} \pi(a|s) \log\left(\frac{\pi(a|s)}{\tilde{\pi}(a|s)}\right) \end{aligned} \tag{6.26}$$

A proof of the lower bound on expected discounted reward $J(\tilde{\pi})$ is shown in equation 6.26. We use the fact that $E_{a \sim \pi} A^{\pi}(s, a) = 0$. Let us consider a policy $\tilde{\pi}$ of the form $\tilde{\pi} = (1-\alpha)\pi + \alpha\pi'$. This implies that $P(a \neq \tilde{a}) = \alpha$. In the last step, $\epsilon$ has been used to denote $\alpha^2$, and $D_{KL}$ denotes the Kullback-Leibler divergence.

$$
\begin{aligned}
E_{a \sim \tilde{\pi},}\left[\sum_{t=0}^{\infty}\gamma^{t} A^{\pi}\left(s_{t},a_{t}\right)\right] &= E_{\tilde{a} \sim \tilde{\pi}, a \sim \pi}\left[\sum_{t=0}^{\infty}\gamma^{t}\left(A^{\pi}(s_{t},\tilde{a}_{t})-A^{\pi}(s_{t},a_{t})\right)\right] \\
&= P(a \neq \tilde{a})E_{\tilde{a} \sim \tilde{\pi}, a \sim \pi}\left[\sum_{t=0}^{\infty}\gamma^{t}\left(A^{\pi}(s_{t},\tilde{a}_{t})-A^{\pi}(s_{t},a_{t})\right)\right] \\
&\leq P(a \neq \tilde{a})E_{a \sim \pi}\left[\sum_{t=0}^{\infty}\gamma^{t}\left(A^{\pi}(s_{t},a_{t})+A^{\pi}(s_{t},a_{t})\right)\right] \\
&\leq P(a \neq \tilde{a})E_{a \sim \pi}\left[\sum_{t=0}^{\infty}\gamma^{t}\,2\max_{s} A^{\pi}(s,a)\right] \\
&= P(a \neq \tilde{a})\frac{2}{1-r}\max_{s,a} A^{\pi}(s,a) \\
&= \frac{2a}{1-r}\max_{s}\max_{a} A^{\pi}(s,a) \\
&\leq \frac{4r\alpha^{2}}{(1-r)^{2}}\max_{s} E_{a \sim \tilde{\pi}}\,|A^{\pi}(s,a)| \\
&= \frac{4\varepsilon\gamma}{(1-r)^{2}}D_{KL}^{\max}(\pi(.|s)\|\tilde{\pi}(.|s))
\end{aligned}
\tag{6.27}
$$

This gives the policy iteration algorithm used by trust region policy optimization, shown in pseudo-code 14.

Schulman et al. noted that the algorithm gives small step sizes. In order to overcome the problem of small step sizes and to make the algorithm more adaptable to practical applications, they introduced several simplifications. They parameterized the policy by $\theta$. Let us denote the original policy as $\pi_{\theta_{old}}$ and the improved policy as $\pi_{\theta}$:

---

**Algorithm 14**    Trust Region Policy Optimization Algorithm

**Require:** Discount factor $\gamma$, initial policy estimate $\pi_0$

1: **for** each i = 0, 1, 2, ... until convergence **do**

2:     Calculate advantage values $A^{\pi_i}(s,a)$ for all $s \in \mathcal{S}$ and $a \in \mathcal{A}$.

3:     Calculate $L^{\pi_i}(\pi) = J(\pi_i) + \sum_{s \in \mathcal{S}} \rho^{\pi_i}(s) \sum_{a \in \mathcal{A}} \pi(a|s) A^{\pi_i}(s,a)$.

4:     Calculate $C = \dfrac{4\epsilon\gamma}{(1-\gamma)^2}$.

5:     Calculate $D_{KL}^{\max}(\pi_i, \pi)$ using equation 6.28.

$$D_{KL}^{\max}(\pi_i, \pi) = \max_s \sum_{a \in \mathcal{A}} \pi_i(a|s) \log\left(\frac{\pi_i(a|s)}{\pi(a|s)}\right) \qquad (6.28)$$

6:     Calculate the improved policy for the next iteration using
       equation $\pi_{i+1} = \text{argmax}_\pi \left( L^{\pi_i}(\pi) - C D_{KL}^{\max}(\pi_i, \pi) \right)$.

7: **end for**

---

1.  The point-wise condition on KL divergence is
    replaced by an average condition as shown in
    equation 6.29:

$$\max_\theta L^{\theta_{old}}(\theta) = \max_\theta \sum_{s \in \mathcal{S}} \rho_{\theta_{old}} \sum_{a \in \mathcal{A}} \pi_\theta(a|s) A^{\theta_{old}}(s,a)$$
$$\text{subject to } \bar{D}_{KL}(\theta_{old}, \theta) \leq \delta \qquad (6.29)$$

2.  Discounted state visitation frequency $\rho_{\theta_{old}}$ is
    approximated as shown in equation 6.30:

$$\left(1 + \gamma_\gamma^2 + ...\right)\bar{\rho} = \frac{1}{1-\gamma} \quad \bar{\rho} = \frac{1}{1-\gamma} E_{s \sim \rho_{\theta_{old}}} ... \qquad (6.30)$$

3. The importance sampling function is introduced in estimation of the objective function. This function should have support over the entire range of values spanned by $\pi_\theta$.

4. The advantage function is rewritten as $Q^{\theta_{old}}(s,a) - V^{\theta_{old}}(s)$. Since $V^{\theta_{old}}(s)$ is not a function of $\theta$, it drops out of the objective function.

5. Finally, the optimization problem of the algorithm is reformulated as shown in equation 6.31:

$$\max_{\theta} E_{s \sim \rho_{\theta_{old}}, a \sim q} \left[ \frac{\pi_\theta(a|s)}{q(a|s)} Q_{\theta_{old}}(s,a) \right]$$

$$\text{subject to } E_{s \sim \rho_{\theta_{old}}} \left[ \bar{D}_{KL}(\pi_{\theta_{old}}(.|s) \ \pi_\theta(.|s)) \right] \leq \delta$$

(6.31)

6. $Q_{\theta_{old}}(s,a)$ is evaluated in a Monte Carlo framework on a stochastically sampled path. This is done using a single path or by sampling multiple paths and selecting a subset of states along those paths. On each state within that subset, an $n$-step rollout of policy is performed. This latter method is called the vine procedure.

7. Approximate the KL divergence constraint as a quadratic function $\bar{D}_{KL} \approx \frac{1}{2}(\theta - \theta_{old})^T A(\theta - \theta_{old})$ where $A$ is the Fisher information matrix computed as shown in equation 6.32:

$$\bar{D}_{KL} \approx \frac{1}{2}\left(\theta - \theta_{old}\right)^{T} A\left(\theta - \theta_{old}\right)$$

$$A_{i,j} = \frac{\partial}{\partial\theta_{i}} \frac{\partial}{\partial\theta_{j}} D_{KL}\left(\theta_{old}, \theta\right)$$

$$= \frac{\partial}{\partial\theta_{i}} \frac{\partial}{\partial\theta_{j}} \pi_{\theta_{old}} \ln\left(\frac{\pi_{\theta_{old}}}{\pi_{\theta}}\right)$$

$$= -\frac{\partial}{\partial\theta_{i}} \frac{\partial}{\partial\theta_{j}} \pi_{\theta_{old}} \ln\pi_{\theta} \qquad (6.32)$$

$$= \frac{\pi_{\theta_{old}}}{\pi_{\theta}^{2}} \frac{\partial\pi_{\theta}}{\partial\theta}^{T} \frac{\partial\pi_{\theta}}{\partial\theta} - \frac{\pi_{\theta_{old}}}{\pi_{\theta}} \frac{\partial}{\partial\theta_{i}} \frac{\partial}{\partial\theta_{j}} \pi_{\theta}$$

$$\approx \frac{\pi_{\theta_{old}}}{\pi_{\theta}^{2}} J^{T} J \text{ where J denotes the Jacobian matrix}$$

8. Finally, using the conjugate gradient method, equation 6.33 is solved to give the search direction $\theta - \theta_{old}$. $J$ denotes the Jacobian matrix computed using backpropagation:

$$J\left(\theta - \theta_{old}\right) = \sqrt{\delta\pi_{\theta_{old}}} \qquad (6.33)$$

9. Once search direction $\kappa$ is known, a few points along this search direction are used to pick a new value of $\theta$. The value of $\theta$ from this set that maximizes the objective function is selected, as shown in equation 6.34:

$$\theta_{i} = \theta_{old} + \alpha_{i}\kappa \text{ for } i = 0,1,\ldots$$

$$\theta = \underset{\theta_{i}}{\text{argmax}} \, E_{s \sim \rho_{\theta_{old}}, a \sim q}\left[\frac{\pi_{\theta_{i}}(a|s)}{q(a|s)} Q_{\theta_{old}}(s,a)\right] \qquad (6.34)$$

# 6.7 Natural Actor-Critic Algorithm: NAC

The natural actor-critic reinforcement learning algorithm was first proposed by Konda and Tsitsiklis (1999) in a paper titled "Actor-Critic Algorithms" and further discussed by Kakade (2001) in his work "A Natural Policy Gradient." The policy gradient algorithm often gets stuck in local maxima, and its speed of convergence near a local optimum is linear. This is because gradient descent takes locally optimal actions and has to compensate for this myopic behavior by taking small steps. Near an optimum, one should take steps given by Newton's method as shown in equation 6.35. The natural actor-critic algorithm is inspired by this approach for calculating policy gradient.

$$\theta_{new} \leftarrow \theta_{old} - \frac{\nabla_\theta f(\theta)}{\nabla_{\theta\theta} f(\theta)} \tag{6.35}$$

Policy gradient optimizes the expected discounted reward obtained by following a policy. The parameter update rule obtained by applying gradient ascent on this objective function is shown in equation 6.36:

$$\max_\theta J(\pi) = E_{s_0 \sim \rho_0, a_t \sim \pi_\theta, s_t \sim P(s|s_{t-1}, a_t)} \left[ \gamma^t r(s_t) \right]$$
$$\theta_{new} \leftarrow \theta_{old} + \alpha \nabla_\theta J(\pi_\theta) \tag{6.36}$$
$$\nabla_\theta J(\pi_\theta) = \sum_{s \in S} d^\pi(s) \sum_{a \in A} \nabla_\theta \pi(a|s, \theta)(Q^\pi(s,a) - V^\pi(s))$$

The compatible value function satisfies the condition shown in equation 6.37. Using a compatible value function, gradient of discounted reward $J(\pi_\theta)$ can be written as shown in equation 6.38:

$$Q^\pi(s,a) - V^\pi(s) = \nabla_\theta \log(\pi(a|s, \theta))w \tag{6.37}$$

$$\nabla_\theta J(\pi_\theta) = \sum_{s \in S} d^\pi(s) \sum_{a \in A} \nabla_\theta \pi(a|s, \theta) \nabla_\theta \log(\pi(a|s, \theta))w$$
$$= E_{s \sim \rho, a \sim \pi} \left[ \nabla_\theta \log(\pi(a|s, \theta)) \nabla_\theta \log(\pi(a|s, \theta))^T w \right] \tag{6.38}$$

Natural policy gradient uses the inverse of Hessian to multiply with Jacobian matrix, analogous to the Newton update step in equation 6.35. Using a compatible value function, this simplifies to the correction shown in equation 6.39:

$$\theta_{new} \leftarrow \theta_{old} + \alpha \mathbf{G}^{-1}(\theta)\nabla_\theta J(\pi_\theta)$$
$$\mathbf{G}(\theta) = \nabla_{\theta\theta} J(\pi_\theta) = \nabla_\theta \log(\pi(a|s,\theta))\nabla_\theta \log(\pi(a|s,\theta))^T \qquad (6.39)$$
$$\theta_{new} \leftarrow \theta_{old} + \alpha w$$

# 6.8 Proximal Policy Optimization: PPO

Proximal policy optimization was proposed by J. Schulman et al. in 2017 in a paper titled "Proximal Policy Optimization Algorithms." This algorithm builds on the framework of trust region policy optimization by modifying the objective function in a two-fold attempt to simplify the model implementation and address the problem of small step sizes. Proximal policy optimization simplifies the objective function used in TRPO by using a clipped surrogate objective function. Motivation behind using a clipped surrogate function is to remove an approximation made in TRPO of optimizing the objective function and satisfying the constraint separately. The unconstrained optimization that should be solved in TRPO is shown in equation 6.40:

$$\max_\theta E\left[\frac{\pi_\theta(a|s)}{\pi_{\theta_{old}}(a|s)}A - \beta D_{KL}\left(\pi_{\theta_{old}} \| \pi_\theta\right)\right] \qquad (6.40)$$

Denoting $\dfrac{\pi_\theta(a|s)}{\pi_{\theta_{old}}(a|s)}$ by $r(\theta)$, it can be seen that $r(\theta_{old}) = 1$. The constraint on KL divergence penalizes changes to $\theta$ that move $r(\theta)$ away from one. To replicate this feature of the constraint, the objective function is modified by introducing a clip function that first clips $r(\theta)$ to a range between $[1 - \epsilon, 1 + \epsilon]$ and then takes a minimum of unclipped $r(\theta)$ and the clipped value of $r(\theta)$. This is shown in equation 6.41. The PPO algorithm optimizes the function shown in equation 6.41, with no other constraint. This simplifies the implementation, and a judicious choice of $\epsilon$ gives a faster learning rate than TRPO:

$$
\begin{aligned}
\max_{\theta} L^{clipped}(\theta) &= E\Big[\big(\min(r(\theta),clip_{r(\theta)}(1-\epsilon,1+\epsilon))\big)A\Big] \\
&= E\Big[\big(\min(r(\theta),\max(r(\theta),1+\epsilon))\big)A\Big]
\end{aligned}
\tag{6.41}
$$

# 6.9 Deep Deterministic Policy Gradient: DDPG

Deep deterministic policy gradient was proposed by Lillicrap et al. (2016) in a paper titled "Continuous Control with Deep Reinforcement Learning." This algorithm adapts DQN (deep Q-network) to continuous action spaces by using deterministic policy gradient formulated by D. Silver et al. It has the following salient features:

1.  It applies deterministic policy gradient in the actor to maximize the expected discounted rewards, as shown in equation 6.42:

$$
\nabla_{\theta}J(\pi_{\theta}) = E_{s\sim d^{\pi},a\sim\pi_{\theta}}\Big[\nabla_{\theta}\pi_{\theta}(s)\nabla_{\pi}Q^{\pi_{\theta}}(s,\pi_{\theta}(s))\Big]
\tag{6.42}
$$

2.  Like DPG, this is an off-policy algorithm. However, unlike DPG, the action space is continuous, and this renders the use of an $\epsilon$-greedy policy intractable. An $\epsilon$-greedy policy will have to resort to numerical optimization of the action function at each state to arrive at an optimal action, which is intractable for highly nonlinear action functions such as deep neural networks. To overcome this problem and to ensure adequate exploration of action space, DDPG

uses an exploration policy $\mu'(s|\theta_\mu)$ derived by adding a noise sampled from the Ornstein-Uhlenbeck process to the policy being learned, $\mu(s|\theta_\mu)$. This is shown in equation 6.43. $N(m, v)$ denotes the normal distribution with mean $m$ and variance $v$:

$$\mu'\left(s_t|\theta_\mu\right) = \mu\left(s_t|\theta_\mu\right) + \mathcal{N}\left(s_t\right)$$

$$\mathcal{N}\left(s_t\right) \sim N\left(s_t, \frac{\sigma^2}{2M}\left(1-e^{-2M}\right)\right) \qquad (6.43)$$

$$M = 0.15$$

$$\sigma = 0.2$$

3. Like DPG, DDPG uses a replay buffer to sample mini-batches for training.

4. Like DPG, this algorithm uses target and learned parameters for both the actor and critic networks.

5. Unlike DPG, DDPG uses soft target updates instead of copying the learned parameters to target parameters after a certain number of iterations.

The complete DDPG algorithm is shown in pseudo-code 15.

---

**Algorithm 15**    Deep Deterministic Policy Gradient Algorithm

**Require:** Discount factor $\gamma$, soft update parameter $\tau$, initial critic network $Q(s, a|\theta^Q)$, initial actor policy network $\mu(s|\theta^\mu)$, and minibatch size $N$

1: Initialize the target network for critic $Q'$ and actor $\mu'$ with weights $\theta^{Q'} \leftarrow \theta^Q$ and $\theta^{\mu'} \leftarrow \theta^\mu$.

2: Initialize replay buffer $R$.

3: **for** each episode = 1, M **do**

4:      Get initial state $s_0$.

5:      **for** each t = 0,1,2,…,T **do**

6:           Select action $a_t = \mu(s_t|\theta^\mu) + \mathcal{N}(s_t)$ where $\mathcal{N}(s_t)$ is an OU process, as shown in equation 6.43.

7:           Take action $a_t$, get reward $r_t$, and transition to next state $s_{t+1}$.

8:           Store transition $(s_t, a_t, r_t, s_{t+1})$ in replay buffer $R$.

9:             Sample a random mini-batch of $N$ transitions from the replay buffer $R$.

10:            Set the target for the critic to be $y_i = r_i + \gamma Q'\left(s_{i-1}, \mu'(s_{i+1}|\theta^{\mu'})|\theta^{Q'}\right)$.

11:            Update the critic network parameters $\theta^Q$ by minimizing the loss function in equation 6.44:

$$\frac{1}{N}\sum_i \left(y_i - Q\left(s_i, a_i|\theta^Q\right)\right)^2 \tag{6.44}$$

12:            Update the actor network parameters by using deterministic policy gradient shown in equation 6.45:

$$\nabla_\theta J(\pi_\theta) = \frac{1}{N}\sum_{i=0}^{N-1}\nabla_\theta \mu^\theta\left(s_i|\theta^\mu\right)\nabla_a Q\left(s_i, a = \mu(s_i|\theta^\mu)|\theta^Q\right) \tag{6.45}$$

13:            Update the target network parameters using a soft update rule shown in equation 6.46:

$$\begin{aligned}
\theta^{Q'} &\leftarrow \tau\theta^Q + (1-\tau)\theta^{Q'} \\
\theta^{\mu'} &\leftarrow \tau\theta^\mu + (1-\tau)\theta^{\mu'}
\end{aligned} \tag{6.46}$$

14:   **end for**

15: **end for**

---

# 6.10 D4PG

Distributed distributional deep deterministic policy gradient, or D4PG, is an enhancement to the deep deterministic policy gradient algorithm. It was proposed by G. Barth-Maron et al. in a paper titled "Distributed Distributional Deterministic Policy Gradients" in 2018. Modifications to DDPG introduced in D4PG are listed in the following:

1.  The state-action value function evaluated by the critic is converted to a distributional form. The output from the critic (state-action value function) is fed to another output layer that produces the parameters of a distribution as an output. To understand this, let us denote $Z_w^\pi \left( Q^\pi \left( s,a \right) \right)$ to be a distribution that takes the output of the critic (i.e., a state-action value function evaluated at state $s$ and action $a$) and produces a distribution with parameters $w$ as output. Hence, $Z_w^\pi$ maps the set of real numbers to a distribution with parameters $w$.

2.  The authors considered three parameterized distributions $Z_w^\pi$: categorical distribution, Gaussian mixture distribution, and a scalar value. Categorical distribution consists of a set of $N$ weights $\omega_i$ with $i = 0, 1, \cdots, N - 1$. The state-action value $Q^\pi(s, a)$ is assumed to lie between $\left[ Q_{min}, Q_{max} \right]$, and this range

373

is divided into $N$ intervals with $\Delta = \dfrac{Q_{max} - Q_{min}}{N-1}$.
Probability that $Z = Q_i = Q_{min} + i\Delta$ is given by
equation 6.47:

$$Z = Q_i = Q_{min} + i\Delta \text{ with probability} \propto \exp(\omega_i) \tag{6.47}$$

Under the categorical distribution, the distance
between the two distributions is defined using cross
entropy loss as shown in equation 6.48. If the output
$Z_w$ does not lie between $[Q_{min}, Q_{max}]$, the hat function
projection shown in equation 6.48 is applied to $Z_w$ to
obtain $Z_w^{proj}$:

$$d(Z, Z_w) = \sum_{i=0}^{N-1} p_i \frac{\exp(\omega_i)}{\sum_{j=0}^{N-1} \exp(\omega_j)} \tag{6.48}$$

$$Z_w^{proj} = \begin{cases} 1 & \text{for } Z_w \leq Q_{min}, i = 0 \\[2mm] \dfrac{Z_w - Z_{i-1}}{Z_i - Z_{i-1}} & \text{for } Z_{i-1} \leq Z_w < Z_i \\[2mm] \dfrac{Z_{i+1} - Z_w}{Z_{i+1} - Z_i} & \text{for } Z_i \leq Z_w < Z_{i+1} \\[2mm] 1 & \text{for } Z_w \geq Q_{max}, i = N-1 \end{cases}$$

A mixture of Gaussians distribution considers a set of
$N$ Gaussians with parameters $w_i = (\omega_i, \mu_i, \sigma_i^2)$ denoting
the weight, mean, and variance of the Gaussian
component $i$. The probability of $Z = Q_l$ is given by
equation 6.49:

$$p(z) \propto \sum_{i=0}^{N-1} \omega_i N(z|\mu_i, \sigma_i^2) \tag{6.49}$$

The distance between the two distributions is defined using KL divergence, as shown in equation 6.50. Since $z$ is a deterministic distribution and $z_w$ is a mixture of Gaussian distribution, only the part corresponding to a non-zero value of $z$ is retained. The expression is evaluated using stochastically selected paths $(s_t, a_t, r_t, s_{t+1})$. $z_j = Q(s_j, a_j)$ denotes the state-action value:

$$
\begin{aligned}
d(Z, Z_w) &= \sum_j z \log\left(\frac{z_w}{z}\right) \\
&= \sum_j \log\left(p(z_j)\right) \\
&= \sum_j \log\left(p(r_j + \gamma z_{j+1})\right) \\
&= \sum_j \log\left(\sum_{i=0}^{N-1} \frac{\omega_i N\left(r_j + \gamma z_{j+1}, \mu_i, \sigma_i^2\right)}{\sum_{k=0}^{N-1} \omega_k N\left(r_j + \gamma z_{j+1} \mid \mu_k, \sigma_k^2\right)}\right)
\end{aligned}
\tag{6.50}
$$

Scalar value distribution is an identity distribution, equivalent to applying no transformation to the input. This corresponds to using the output of the critic as the state-action value function. The distance measure between $Z$ and $Z_w$ in this case is the mean square loss function. Using the scalar distribution function is equivalent to not using distributional form, giving distributed deep deterministic policy gradient or D3PG.

3. The loss function minimized by the critic takes the form shown in equation 6.48 or 6.50. The discounted reward function maximized by the actor takes the form shown in equation 6.51:

$$
J(\theta) = E\left[\nabla_\theta \pi_\theta(s) E\left[\nabla_a Z_w(s, a = \pi_\theta(s))\right]\right]
\tag{6.51}
$$

4. TD error is estimated using $n$-step update in place of the customary TD0 update employed in DDPG.

5. $K$ actors explore the state-action space, adding experiences to the replay buffer in parallel. This step distributes the process of gathering experience among $K$ actors, accounting for the "distributed" term in the D4PG acronym.

6. D4PG uses a prioritized replay buffer as described by T. Schaul et al. in their paper titled "Prioritized Experience Replay." Items are sampled from the replay buffer with probability of selecting element $i$ given by equation 6.52 . rank($i$) is the rank of experience $i$ when sorted in descending order by TD error, $\delta_i$. Another version of the prioritized replay buffer sets $p_i = |\delta_i| + \epsilon$:

$$P(i) = \frac{p_i^\alpha}{\sum_j p_j^\alpha}$$

$$p_i = \frac{1}{\text{rank}(i)}$$

(6.52)

The complete D4PG algorithm is sketched in pseudo-code 16.

## 6.11 TD3PG

Twin delayed deep deterministic policy gradient (TD3PG) was introduced in 2018 by S. Fujimoto et al. in a paper titled "Addressing Function Approximation Error in Actor-Critic Methods." The algorithm is an enhancement to DDPG for dealing with the overestimation bias induced by function approximation in the critic coupled with parallel policy learning in the actor, leading to high variance and, occasionally, divergence. The analogous method for handling the overestimation bias in Q-learning is DDQN (double deep Q-network), which uses a target network to minimize propagation of overestimation errors. TD3PG can be viewed as a similar approach to deal with the overestimation bias in the context of deterministic policies in continuous space using actor-critic methods.

The TD3PG algorithm incorporates the following modifications to DDPG in order to address the problem of the overestimation bias and variance reduction:

---

**Algorithm 16**    Distributed Distributional Deep Deterministic Policy Gradient Algorithm

**Require:** Discount factor $\gamma$, initial critic network $Q(s, a| \theta^Q)$, initial actor policy network $\mu(s| \theta^\mu)$, mini-batch size $M$, trajectory length $N$, number of actors $K$, replay buffer size $R$, exploration constant $\epsilon$, $t_{target}$ time period for updating target network parameters, $t_{actor}$ time period for replicating parameters to actors, initial learning rates $\alpha_0$ and $\beta_0$, and an annealing schedule for the learning rates

1: Initialize the target network for critic $Q'$ and actor $\mu'$ with weights
   $\theta^{Q'} \leftarrow \theta^Q$ and $\theta^{\mu'} \leftarrow \theta^\mu$.

2: Launch $K$ actors and replicate network weights for each actor. Each actor samples action $a = \pi_\theta(s) + \sigma\epsilon$ with $\epsilon \sim N(0,1)$. Execute action $a$, obtain reward $r$, transition to state $s'$, and store the experience in the replay buffer.

3: **for** each $t = 0,1,2,...,T$ **do**

4:    Sample $M$ transitions each of size $N$ from the replay buffer.

5:    Construct the target distributions as shown in equation 6.53:

$$Y_i = \sum_{n=0}^{N-1} \gamma^n r_{i+n} + \gamma^N Z_{w'}\left(s_{i+N}, \pi_{\theta'}\left(s_{i+N}\right)\right) \tag{6.53}$$

6:    Calculate the actor and critic updates using equation 6.54:

$$\delta_w = \frac{1}{M}\sum \nabla_w \text{Loss}$$

$$\delta_\theta = \frac{1}{M}\sum_i \nabla_\theta \pi_\theta\left(s_i\right) E\left[\nabla_a Z_w(s_i, a = \pi_\theta\left(s_i\right))\right] \tag{6.54}$$

7:    Update target networks from learned parameters after $t_{target}$ timesteps, that is, if $t = 0 \bmod t_{target}$ .

8:    If $t = 0 \bmod t_{actor}$ , replicate network parameters to actors.

9: **end for**

---

1.  A clipped double Q-network is used for learning the value function in the critic. The approach followed by DDQN to address the overestimation problem by using separate target and learned networks is found to be insufficient in an actor-critic context because the policy changes slowly causing the target and learned networks in the critic to become similar. In order to address this problem, TD3PG uses a single policy $\pi_\phi$ and two state-action value functions in the critic, $Q_{\theta_1}$ and $Q_{\theta_2}$. Like DDQN, the corresponding networks used for calculating the target value are $\pi_{\phi'}$, $Q_{\theta_1'}$, and $Q_{\theta_2'}$. Targets used for learning the value function are constructed using the minimum value over the two target value functions as shown in equation 6.55. It should be noted that both the targets use the same policy $\pi_{\phi'}$:

$$y_1 = y_2 = r + \gamma \min_{i=1,2} Q_{\theta_i'}\left(s', \pi_{\phi'}(s')\right) \tag{6.55}$$

2. To address the problem of high variance, the TD3PG algorithm introduces two steps of delaying updates to the target network. These two steps account for the name "twin delayed" in the algorithm's name. In the first step introducing a delay, actor parameters are updated only once every $d$ time steps. This delay is added to learn the policy only after the value function in the critic has undergone a certain number of corrections. In the second delay, the target function parameters are updated gradually using a parameter $\tau << 1$, as shown in equation 6.56:

$$\begin{aligned}\theta_i' &\leftarrow \tau\theta_i + (1-\tau)\theta_i' \text{ for i} = 1,2 \\ \phi' &\leftarrow \tau\phi + (1-\tau)\phi'\end{aligned} \tag{6.56}$$

3. The actor adds noise while exploring the action space.

The complete TD3PG algorithm is sketched in pseudo-code 17.

# 6.12 Soft Actor-Critic: SAC

The soft actor-critic algorithm was proposed by T. Haarnoja et al. in a paper titled "Soft Actor-Critic: Off-Policy Maximum Entropy Deep Reinforcement Learning with a Stochastic Actor" in 2018. The algorithm seeks to address the problem of hyper-parameter non-generalization in model-free deep reinforcement learning algorithms applied to different problems. Problem-specific hyper-parameter selection is necessitated by the large size of the training dataset and non-robust convergence properties of the algorithm. The algorithm's convergence is intimately tied to the size of the training dataset, and meticulous selection of hyper-parameters is called for. The values of hyper-parameters differ markedly across different problems, restricting the applicability of model-free deep reinforcement learning algorithms.

---

**Algorithm 17**    Twin Delayed Deep Deterministic Policy Gradient Algorithm

**Require:** Discount factor $\gamma$, initial critic networks $Q_1(s, a|\theta_1)$, $Q_2(s, a|\theta_1)$, initial actor policy network $\pi(s|\phi)$, mini-batch size $M$, $d$ time period for updating target network parameters, $\tau$ soft update weight, learning rates $\alpha$ and $\beta$.

1: Initialize the target networks for critic and actor with weights
   $\theta_1' \leftarrow \theta_1$, $\theta_2' \leftarrow \theta_2$, and $\phi' \leftarrow \phi$.

2: Initialize the replay buffer.

3: **for** each t = 0,1,2,...,T **do**

4:    Select an action $a \sim \pi_\phi(s) + \epsilon$ with exploration noise
      $\epsilon \sim N(0, \sigma^2)$.

5:    Observe reward, transition to the next state, and store the experience in replay buffer.

6:    Sample a mini-batch of $M$ transitions $(s, a, r, s')$ from the replay buffer.

7:    Calculate the action $\tilde{a} \leftarrow \pi_{\phi'}(s') + \epsilon$  where
      $\epsilon \sim clip(N(0, \sigma^2), -c, c)$.

8:    Calculate the target value for training $Q_1$ and $Q_2$ as shown in equation 6.57:

$$y = r + \gamma \min_{i=1,2} Q_{\theta_i'}(s', \tilde{a})$$ (6.57)

9:    Update learned networks in the critic using equation 6.58:

$$\theta_i \leftarrow \underset{\theta_i, i=1,2}{\mathrm{argmin}} \frac{\Sigma(y - Q_i(s, a | \theta_i))^2}{M}$$ (6.58)

10:   If $t = 0 \bmod d$, perform a delayed update:

1. Update $\phi$ using deterministic policy gradient on $Q_1$ as shown in equation 6.59:

$$\nabla_\phi J(\pi_\phi) = \frac{\Sigma \nabla_a Q_1(s, a) \nabla_\phi \pi_\phi(s)}{M}$$ (6.59)

2. Perform a soft update on target network parameters using equation 6.56.

11: **end for**

---

The soft actor-critic algorithm augments the reward function by adding entropy of the policy function, as shown in equation 6.60. It seeks to maximize the total expected rewards and also maximize the policy entropy. Maximizing the policy entropy means that the algorithm is not forced to pick an arbitrary action from a set of actions that produce similar rewards in order to converge. Maximizing total rewards ensures that actions that yield greater total (discounted) rewards are selected by the policy over less promising actions. Due to the entropy term (logarithm of policy), the soft actor-critic algorithm is applicable to continuous action spaces:

$$\max_{\pi} J^{\pi} = \max_{\pi}\left[ r\left(s_t,a_t,s_{t+1}\right)+\gamma r\left(s_{t+1},a_{t+1},s_{t+2}\right)+\cdots-\log\left(\pi\left(\cdot|s_t\right)\right)\right]$$
$$= \max_{\pi}\left[ R\left(s_t,a_t,s_{t+1}\right)-\log\left(\pi\left(\cdot|s_t\right)\right)\right] \tag{6.60}$$
$$= \max_{\pi}\left[ Q^{\pi}\left(s_t,a_t\right)-\log\left(\pi\left(\cdot|s_t\right)\right)\right]$$

The algorithm uses a state value function, $V_{\psi}(s_t)$, and a state-action value function, $Q_{\theta}(s_t, a_t)$. The critic minimizes the state value function using 6.61 and the state-action value function using equation 6.62. As can be seen from equations 6.61 and 6.62, training the state value function requires the state-action value function, and training the state-action value function requires the state value function:

$$\min_{\psi} J_V\left(\psi\right)=\frac{1}{2}E\left[\left(V_{\psi}\left(s_t\right)-E_{a_t\sim\pi}(Q_{\theta}(s_t,a_t)-\log\pi(a_t|s_t))\right)^2\right] \tag{6.61}$$
$$\nabla_{\psi} J_V\left(\psi\right)=\nabla_{\psi}V_{\psi}\left(s_t\right)\left(V_{\psi}\left(s_t\right)-E_{a_t\sim\pi}(Q_{\theta}(s_t,a_t)-\log\pi(a_t|s_t))\right)$$

$$\min_{\theta} J_Q\left(\theta\right)=\frac{1}{2}E\left[\left(Q_{\theta}\left(s_t,a_t\right)-(r(s_t,a_t)+\gamma V_{\tilde{\psi}}\left(s_{t+1}\right))\right)^2\right] \tag{6.62}$$
$$\nabla_{\theta} J_Q\left(\theta\right)=\nabla_{\theta}Q_{\theta}\left(s_t,a_t\right)\left(Q_{\theta}\left(s_t,a_t\right)-r(s_t,a_t)-\gamma V_{\tilde{\psi}}\left(s_{t+1}\right)\right)$$

In order to stabilize learning using gradient descent, the critic uses a target state value function $V_{\tilde{\psi}(s_t)}$ that is updated gradually using soft update equation 6.63 and a small value of $\tau < < 1$. Further, it uses two state-action value functions, $Q_{\theta_1}\left(s_t,a_t\right)$ and $Q_{\theta_2}\left(s_t,a_t\right)$, that are trained independently. $Q(s_t, a_t)$ is then set to be minimum of $Q_{\theta_1}\left(s_t,a_t\right)$ and $Q_{\theta_2}\left(s_t,a_t\right)$:

$$\tilde{\psi} \leftarrow \tau\psi +\left(1-\tau\right)\tilde{\psi} \text{ where } \tau << 1 \tag{6.63}$$

The policy is updated by minimizing the KL divergence between the policy function $\pi_{\phi}(a_t|s_t)$ and $\dfrac{\exp\left(Q_{\theta}\left(s_t,a_t\right)\right)}{\int_{a_t}\exp\left(Q_{\theta}\left(s_t,a_t\right)\right)da_t}$. Derivation of this term is illustrated in equation 6.64. We are trying to maximize $Q_{\theta}(s_t, a_t)$ –

representing the total discounted rewards – and also maximize the entropy. In a stochastic gradient ascent framework, we sample one or a few actions from the policy. This can be written using the action probability density specified by the policy, as shown in equation 6.64. Finally, the last term corresponds to KL divergence. $Z_\theta(s_t)$ is a normalizing constant equal to $\int_{a_t} \exp\left(Q_\theta\left(s_t, a_t\right)\right) da_t$. Let us assume the policy function is represented as a deep neural network, with $a_t = f(\phi, s_t)$, where $\phi$ represents the parameters of the neural network. Gradient of this term is shown in equation 6.65:

$$\max_\phi E_{a_t \sim \pi\phi}[Q_\theta(s_t, a_t) - \log \pi_\phi(a_t \mid s_t)]$$

$$\max_\phi \pi_\phi(a_t \mid s_t)[Q_\theta(s_t, a_t) - \log \pi_\phi(a_t \mid s_t)]$$

$$\max_\phi \pi_\phi(a_t \mid s_t)[\log(\exp Q_\theta(s_t, a_t)) - \log \pi_\phi(a_t \mid s_t)]$$

$$\max_\phi \pi_\phi(a_t \mid s_t)\left[\log \frac{\exp Q_\theta(s_t, a_t)}{\pi_\phi(a_t \mid s_t)}\right]$$

$$\max_\phi D_{KL}\left[\pi_\phi(a_t \mid s_t) \| \frac{\exp(Q_\theta(s_t, a_t))}{\int_{at} \exp(Q_\theta(s_t, a_t))d_{at}}\right] \qquad (6.64)$$

$$\max_\phi D_{KL}\left[\pi_\phi(a_t \mid s_t) \| \frac{\exp(Q_\theta(s_t, a_t))}{z_\theta(s_t)}\right]$$

$$= \max_\phi J_\pi(\phi)$$

$$= \min_\phi - J_\pi(\phi)$$

$$\nabla_\phi J_\pi(\phi) = \left(\nabla_{at} Q_\phi(s_t, a_t) - \nabla_{at} \log \pi_\phi(a_t \mid s_t)\right)\nabla_\phi f(\phi, s_t) - \nabla_\phi \log \pi_\phi(a_t \mid s_t) \qquad (6.65)$$

Policy training can be stabilized by adding an external source of disturbance, e.g., $\epsilon \sim N(0, 1)$, and constructing a random variable $Z = \mu + \sigma\epsilon$. Action can then be produced as an output from the policy network depending upon the value of $\epsilon$ sampled, as shown in equation 6.66. $\mu$ and $\sigma$ are additional network parameters that are learned using stochastic gradient ascent:

$$a_t = f\left(Z|s_t,\phi,\mu,\sigma\right)$$
$$Z = \mu + \sigma\epsilon$$

(6.66)

The complete soft actor-critic algorithm is sketched in pseudo-code 18.

# 6.13  Variational Autoencoder

Variational autoencoders were first introduced by Kingma and Welling in 2014 in a paper titled "Auto-encoding Variational Bayes" at the International Conference on Learning Representations. A variational autoencoder, abbreviated as VAE, is a generative model that has the ability to map a set of inputs to an underlying probability space of a latent (hidden) variable and to sample from that space to generate new observations. The distinction between generative models (like VAE, GAN) and discriminative models (like CNN, SVM) is a key dichotomy in machine learning. Discriminative models predict an output corresponding to an input. For example, a classifier attempts to classify input data into classes, and a regression model produces an output value. Generative models, on the other hand, produce new samples of data, ostensibly similar to the input data. In this sense, generative models learn the underlying probability distribution of data in order to draw new samples from it. Discriminative models, on the other hand, do not need to learn underlying probability density governing the data distribution; it is sufficient for them to identify certain discriminating features in order to classify input data (classifier) or produce an output value (regressor).

At the outset, modeling the probability distribution space of input data seems like a daunting task: not just because of multidimensional data but also because of unavailability of a tool to model the underlying distribution space.

A VAE learns a stochastic mapping between observed data, which is sampled from an underlying multidimensional probability density, and a latent variable $z$, which is assumed to have a relatively simple, low-dimensional probability density. In this sense, it can be viewed as a tool for mapping a high-dimensional input space to a low-dimensional latent space – a process known as encoding. New samples are generated by sampling from the low-dimensional latent space and applying an inverse mapping to generate an input from the latent variable using a component called a decoder. Deep neural networks are used for encoding the input to a latent variable (encoder) and decoding the latent variable to a reconstructed input (decoder).

---

## Algorithm 18    Soft Actor-Critic Algorithm

**Require:** Discount factor $\gamma$, initial critic networks $V(s|\psi)$, $Q(s, a|\theta)$, initial actor policy network $\pi(a|s, \phi)$, mini-batch size $M$, $\tau$ soft update weight, learning rates $\alpha_V$, $\alpha_Q$, and $\alpha_\pi$.

1: Initialize the replay buffer.

2: Initialize the target network for the critic with weights $\tilde{\psi} \leftarrow \psi$ .

3: Copy the state-action value network to $Q(s, a|\theta_1)$ and $Q(s, a|\theta_2)$.
   $\theta_1 \leftarrow \theta$ and $\theta_2 \leftarrow \theta$.

4: **for** each episode **do**

5:    **for** each t = 0,1,2,...,T **do**

6:       Select an action $a_t \sim \pi_\phi(a_t|s_t)$.

7:       Observe reward $r_t$, transition to the next state $s_{t+1}$, and store the experience $(s_t, a_t, r_t, s_{t+1})$ in the replay buffer.

8:       **for** each i = 0, 1 **do**

9:          Sample a mini-batch of $M$ transitions $(s, a, r, s')$ from the replay buffer.

10:        Calculate the gradients using equations 6.61, 6.62, and
           6.65. Use $Q(s_t, a_t) = \min\left(Q_{\theta_1}(s_t, a_t), Q_{\theta_2}(s_t, a_t)\right)$.

11:        Update the network parameters as shown in
           equation 6.67:

$$
\begin{aligned}
\psi &\leftarrow \psi - \alpha_V \nabla_\psi J_V(\psi) \\
\theta_i &\leftarrow \theta_i - \alpha_Q \nabla_\theta J_Q(\theta_i) \\
\phi &\leftarrow \phi - \alpha_\pi \nabla_\phi J_\pi(\phi) \\
\tilde{\psi} &\leftarrow \tau\psi + (1-\tau)\tilde{\psi}
\end{aligned}
\tag{6.67}
$$

12:    **end for**

13:    **end for**

14: **end for**

---

Let us denote input data by $X$ and assume that the input is generated
by a transformation of a low-dimensional latent variable. Let us denote
the underlying latent space by $z$. Using the Bayes theorem, probability
density of $X$ can be written using posterior density $P(X|z)$, as shown in
equation 6.68:

$$
P(X) = \int P(X|z)P(z)dz = \int P(X,z)dz \tag{6.68}
$$

Let us use a model parameterized by $\theta$ (e.g., a deep neural network) to
learn the mapping of the high-dimensional probability space of an input
variable to the probability space of a low-dimensional latent variable $z$, $P_\theta(z|X)$.
Let us select another model parameterized by $\phi$ for mapping the latent
variable back to the input variable. By the principle of maximum likelihood, we
want to select parameters $\theta$ and $\phi$ that maximize the probability of observing
the data $X$. This is equivalent to maximizing the log probability. Input samples
are drawn from the unknown probability density of the input variable. We can
convert the sum of probabilities over inputs $X$ to an expectation over the latent
variable's probability density. This is shown in equation 6.69:

$$\log \prod P_\theta(X) = \sum \log P_\theta(X) = E_X \log P_\theta(X)$$
$$= E_{q_\phi(z|X)}\left[\log P_\theta(X)\right]$$
$$= E_{q_\phi(z|X)}\left[\log \frac{P_\theta(X,z)}{P_\theta(z|X)}\right]$$
$$= E_{q_\phi(z|X)}\left[\log \frac{P_\theta(X,z)\, q_\phi(z|X)}{q_\phi(z|X)\, P_\theta(z|X)}\right] \qquad (6.69)$$
$$= E_{q_\phi(z|X)}\left[\log \frac{P_\theta(X,z)}{q_\phi(z|X)}\right] + E_{q_\phi(z|X)}\left[\frac{q_\phi(z|X)}{P_\theta(z|X)}\right]$$
$$= \text{ELBO} + D_{KL}\left(q_\phi(z|X)\ \ P_\theta(z|X)\right)$$

In equation 6.69, ELBO denotes evidence lower bound. It is also known as variational lower bound. $D_{KL}$ denotes Kullback-Leibler divergence and is always non-negative. Equation 6.69 implies that $E\left[\log P_\theta(X)\right] \geq$ ELBO. Because of this inequality, maximizing ELBO also maximizes the log likelihood of observing the data. We can rewrite ELBO using the Bayes theorem, as shown in equation 6.70:

$$\text{ELBO} = E_{q_\phi(z|X)}\left[\log \frac{P_\theta(X,z)}{q_\phi(z|X)}\right]$$
$$= E_{q_\phi(z|X)}\left[\log P_\theta(X,z)\right] - E_{q_\phi(z|X)}\left[\log q_\phi(z|X)\right]$$
$$= E_{q_\phi(z|X)}\left[\log P_\theta(X|z)P(z)\right] - E_{q_\phi(z|X)}\left[\log q_\phi(z|X)\right] \qquad (6.70)$$
$$= E_{q_\phi(z|X)}\left[\log P_\theta(X|z)\right] - E_{q_\phi(z|X)}\left[\log q_\phi(z|X) - \log P(z)\right]$$
$$= E_{q_\phi(z|X)}\left[\log P_\theta(X|z)\right] - D_{KL}\left(q_\phi(z|X)\ \ P(z)\right)$$
$$= -\text{reconstruction loss from decoder} - \text{regularization loss}$$

In equation 6.70, $E_{q_\phi(z|X)}\left[\log P_\theta(X|z)\right]$ represents the negative of reconstruction loss from the decoder for reconstructing input $X$ using the latent variable $z$. In addition, we also want the modeled probability density of latent variable $z$, $q_\phi(z|X)$, to be as close as possible to the unknown, true probability density of $z$, $P(z)$. We want this probability to depend upon input data samples; hence, we select posterior probability $q_\phi(z|X)$. This term can be viewed as regularization loss. Finally, putting the expression of ELBO from equation 6.70 back into equation 6.69, we get equation 6.71:

$$E_X \log P_\theta(X) - D_{KL}\left(q_\phi(z|X)\|P_\theta(z|X)\right) = E_{q_\phi(z|X)}\left[\log P_\theta(X|z)\right] -$$
$$D_{KL}\left(q_\phi(z|X)\|P(z)\right) \quad (6.71)$$
$$= \text{ELBO}$$

Equation 6.71 states that maximizing ELBO will maximize the probability of observing the data while also minimizing the distance between the modeled probability density and true posterior probability density of the latent variable, that is, between $q_\phi(z|X)$ and $P_\theta(z|X)$. In order to achieve this, we must minimize the reconstruction loss from the decoder while also minimizing the regularization loss.

The encoder network maps multidimensional probability density of input variable $X$ into low-dimensional probability density of latent variable $z$, while the decoder samples from the learned probability density space of $z$, $q_\phi(z|X)$, to reconstruct $X$. Let us assume the prior distribution of latent variable $z$ to be standard normal, i.e., $N(0, 1)$. Further, let us assume we restrict the space of posterior density of the latent variable to normal distributions, i.e., $q_\phi(z|X) = N(\mu(X), \Sigma(X))$. Kullback-Leibler divergence representing the regularization loss term can be simplified as shown in equation 6.72. $K$ represents the dimension of input $X$:

$$D_{KL}\left(q_\phi(z|X)\|P(z)\right) = D_{KL}\left(N(\mu(X), \Sigma(X))\|N(0, I)\right)$$
$$= \frac{1}{2}\left[tr(\Sigma(X)) + \mu(X)^T \mu(X) - K - \log\det(\Sigma(X))\right] \quad (6.72)$$

The final loss function can be written as shown in equation 6.73. The encoder maps the input $X$ to mean $\mu(X)$ and standard deviation $\Sigma(X)$. Using the **reparameterization trick**, we sample a random number from uniform normal distribution $\epsilon \sim N(0, 1)$ and construct a latent variable sample $Z = \mu(X) + \epsilon\Sigma(X)$. The decoder uses this latent variable sample $Z$ to reconstruct an output, $f(Z)$. Reconstruction loss is $\|X - f(Z)\|^2$ from the decoder, while regularization loss is $D_{KL}(N(\mu(X), \Sigma(X))\|N(0, I))$. $\theta$ denotes the parameters of the encoder network, and $\nu$ denotes the parameters of the decoder network. This is depicted in Figure 6-1.

$$\min_{\theta,\nu}\left[\|X - f_\nu(Z)\|^2 + D_{KL}\left(N(\mu_\theta(X), \Sigma_\theta(X))\|N(0,I)\right)\right] \qquad (6.73)$$

# 6.14  VAE for Dimensionality Reduction

In this section, let us apply variational autoencoders for dimensionality reduction. Multidimensional data, such as images, is easier to store and classify using dimensionality reduction. The traditional approach for image compression involved the use of SVD and decomposition of an image into eigen-images. However, this approach still required storing eigen-images. Let us use variational autoencoders for reducing image dimensions to six: three dimensions each for storing the projected mean and variance using VAE. Dimensionally reduced images can be used for quick image recognition, for example, at an ATM.

Let us use a Kaggle dataset comprising of 6,899 images from eight distinct categories as shown in Table 6-1.

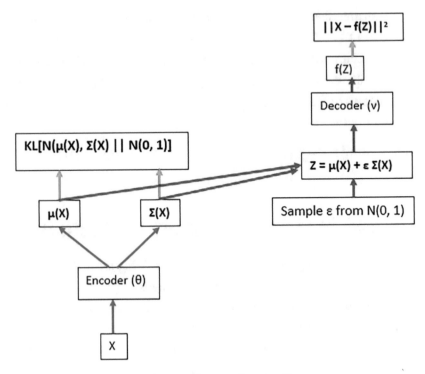

***Figure 6-1.*** *Encoder and Decoder Within VAE*

The model uses the LeakyReLU activation function to ensure that the network does not get saturated for negative activation values. The LeakyReLU activation function has been covered in Chapter 2. The encoder uses a convolutional neural network (CNN) to encode the image into a latent three-dimensional space of a normal distribution defined by mean and variance. After training, images from the testing dataset are projected onto the latent space using the encoder only. In order to visualize the distribution of images in the three-dimensional latent space, pairplots for mean and variance are shown in Figures 6-2 and 6-3.

As seen from the plot in Figure 6-2, images of flower and airplane begin to cluster in distinct segments of the mean space, while the distinction between other objects is not as clear-cut. The log variance pairplot of projected images shows a similar result. On the YZ plane, images begin to cluster in distinct segments on a line. This example shows the ability of a VAE to assign similar images to distinct subsections of a lower-dimensional space and showcases its usefulness for dimensionality reduction.

**Table 6-1.**  *Image Counts in Each Class*

| Object | Count |
| --- | --- |
| airplane | 727 |
| car | 968 |
| cat | 885 |
| dog | 702 |
| flower | 843 |
| fruit | 1000 |
| motorbike | 788 |
| person | 986 |

The code for projecting the images onto a lower-dimensional space using a VAE is shown in Listing 6-2.

**Listing 6-2.**  Projecting Images onto a Lower-Dimensional Space Using a VAE

```
1    import math
2    import os
3
4    import PIL
5    import PIL.Image
6    import matplotlib.pyplot as plt
7    import numpy as np
```

```
8    import pandas as pd
9    import seaborn as sns
10   import tensorflow as tf
11   from tensorflow.keras import layers
12
13   from src.learner.VAE import VariationalAutoEncoder
14
15   tf.random.set_seed(10)
16   np.random.seed(10)
17
18
19   class LeakyRelu(object):
20       def __init__(self, alpha):
21           self.alpha = alpha
22
23       def __call__(self, x):
24           return tf.nn.leaky_relu(x, alpha=self.alpha)
25
26
27   class Encoder(tf.keras.Model):
28       """ Maps mnist digits to (z_mean, z_log_var, z) """
29       def __init__(self, latent_dim, name="encoder",
             alpha=0.1, **kwargs):
30           super(Encoder, self).__init__(name=name,
             **kwargs)
31           self.conv0 = layers.Conv2D(8, 3, padding="same",
             activation="relu")
32           self.conv1 = layers.Conv2D(32, 3, strides=2,
             padding="same", activation=LeakyRelu(alpha))
33           self.conv2 = layers.Conv2D(64, 3, strides=2,
             padding="same", activation=LeakyRelu(alpha))
```

```
34          self.flatten = layers.Flatten()
35          self.dense1 = layers.Dense(16)
36          self.mean = layers.Dense(latent_dim,
            name="z_mean")
37          self.logvar = layers.Dense(latent_dim, name="z_
            log_var")
38
39      def call(self, inputs, **kwargs):
40          x = self.conv1(inputs)
41          x = self.conv2(x)
42          x = self.flatten(x)
43          x = self.dense1(x)
44          z_mean = self.mean(x)
45          z_log_var = self.logvar(x)
46          return z_mean, z_log_var
47
48
49  class Decoder(tf.keras.Model):
50      """ Converts z back to readable digit """
51      def __init__(self, name="decoder", alpha=0.1,
        **kwargs):
52          super(Decoder, self).__init__(name, **kwargs)
53          self.dense1 = layers.Dense(25 * 25 * 64)
54          self.reshape = layers.Reshape((25, 25, 64))
55          self.convt1 = layers.Conv2DTranspose(64,
            3, activation=LeakyRelu(alpha), strides=2,
            padding="same")
56          self.convt2 = layers.Conv2DTranspose(32,
            3, activation=LeakyRelu(alpha), strides=2,
            padding="same")
```

```
57          self.convt3 = layers.Conv2DTranspose(3, 3,
            activation="tanh", padding="same")
58
59      def call(self, inputs, **kwargs):
60          x = self.dense1(inputs)
61          x = self.reshape(x)
62          x = self.convt1(x)
63          x = self.convt2(x)
64          x = self.convt3(x)
65          return x
66
67
68  class VAEImages(object):
69      def __init__(self, input_dir, obj_names, img_
        size=(100, 100), batch_size=100, epochs=30,
70                  validation_split=0.2, latent_dim=2):
71          self.input_dir = input_dir
72          self.obj_names = obj_names
73          self.img_size = img_size
74          self.batch_size = batch_size
75          self.epochs = epochs
76          self.latent_dim = latent_dim
77          self.validation_split = validation_split
78          self.vae = None
79
80      def plotImgs(self):
81          counts = np.zeros(len(self.obj_names),
            dtype=np.int32)
82          nrow = 2
83          fig, axs = plt.subplots(nrow, math.ceil(len(self.
            obj_names) / nrow))
```

```
84              for i, obj in enumerate(self.obj_names):
85                  dname = os.path.join(self.input_dir, obj)
86                  obj_list = os.listdir(dname)
87                  counts[i] = len(obj_list)
88                  rand_img = PIL.Image.open(os.path.join(dname,
                    obj_list[0]))
89                  col, row = divmod(i, nrow)
90                  axs[row, col].imshow(np.array(rand_img))
91                  axs[row, col].set_xticks([])
92                  axs[row, col].set_yticks([])
93                  axs[row, col].set_title(obj)
94          plt.show()
95          df = pd.DataFrame({"Object": self.obj_names,
            "Count": counts})
96          print(df)
97
98      def train_vae(self):
99          encoder = Encoder(self.latent_dim)
100         decoder = Decoder()
101         vae = VariationalAutoEncoder(encoder, decoder,
            from_logits=True, cross_entropy_loss=False, kl_
            loss_weight=0.1)
102         train_dataset = tf.keras.utils.image_dataset_
            from_directory(self.input_dir, image_size=self.
            img_size,
103                             batch_size=self.batch_size,
                                seed=10,
104                             validation_split=self.
                                validation_split,
105                             subset="training")
```

```
106         for batch_num, train_batch in enumerate(train_
            dataset):
107             img_data = train_batch[0].numpy().astype(np.
                float32) / 255.0
108             loss = vae.fit(img_data, epochs=self.epochs)
109             print("Batch %d, final loss: %f" % (batch_
                num+1, loss))
110         self.vae = vae
111
112     def test_vae(self):
113         assert self.vae, "VAE needs to be trained first"
114         valid_dataset = tf.keras.utils.image_dataset_
            from_directory(self.input_dir, image_size=self.
            img_size,
115                             batch_size=self.batch_size,
                                seed=10,
116                             validation_split=self.
                                validation_split,
117                             subset="validation")
118         class_names = valid_dataset.class_names
119         mean_x, mean_y, mean_z = [], [], []
120         lvx, lvy, lvz, label, r_mimg, b_mimg, g_mimg =
            [], [], [], [], [], [], []
121         for batch_num, valid_batch in enumerate(valid_
            dataset):
122             img_data = valid_batch[0].numpy().astype(np.
                float32) / 255.0
123             labels = valid_batch[1]
124             mean, log_var = self.vae.encoder(img_data)
125             mean = mean.numpy()
126             log_var = log_var.numpy()
```

```
127              mean_x.extend(mean[:, 0])
128              mean_y.extend(mean[:, 1])
129              mean_z.extend(mean[:, 2])
130              lvx.extend(log_var[:, 0])
131              lvy.extend(log_var[:, 1])
132              lvz.extend(log_var[:, 2])
133              label.extend([class_names[l] for l in
                 labels])
134
135          df = pd.DataFrame({"Label": label, "Mean(X)":
             mean_x, "Mean(Y)": mean_y, "Mean(Z)": mean_z,
136                          "LogVar(X)": lvx, "LogVar(Y)":
                             lvy, "LogVar(Z)": lvz,
137                          })
138
139          sns.pairplot(data=df[["Mean(X)", "Mean(Y)",
             "Mean(Z)", "Label"]], hue="Label")
140          plt.show()
141
142
143          sns.pairplot(data=df[["LogVar(X)", "LogVar(Y)",
             "LogVar(Z)", "Label"]], hue="Label")
144          plt.show()
145
146
147  if __name__ == "__main__":
148      input_dir = r"C:\prog\cygwin\home\samit_000\RLPy\
         data\kaggle_images\natural_images"
149      objs = ["airplane", "car", "cat", "dog", "flower",
         "fruit", "motorbike", "person"]
150      vae_imgs = VAEImages(input_dir, objs, latent_dim=3)
```

```
151        vae_imgs.plotImgs()
152        vae_imgs.train_vae()
153        vae_imgs.test_vae()
```

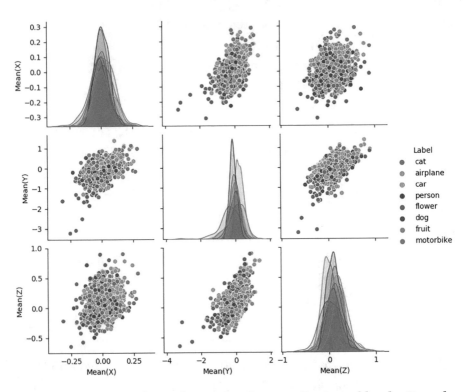

***Figure 6-2.*** *Images from the Testing Dataset Projected by the Encoder on Three-Dimensional Mean Space*

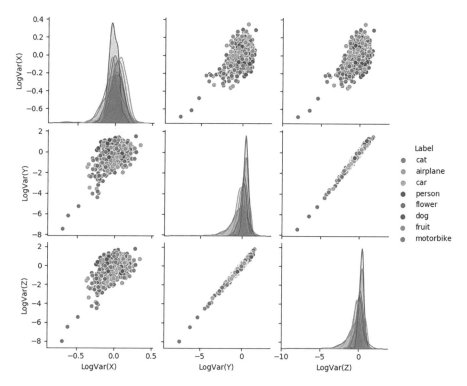

***Figure 6-3.*** *Images from the Testing Dataset Projected by the Encoder on Three-Dimensional Variance Space*

## 6.15 Generative Adversarial Networks

Generative adversarial networks, or GANs, were introduced by Ian Goodfellow et al. in an eponymous paper in 2014. As the name indicates, GANs are generative models that learn the underlying probability distribution of inputs, which can be used to generate new samples from the distribution that are similar to inputs. They are different from VAEs in the methodology adopted for learning the probability distribution of input, $X$. While VAEs use a transformation to map the probability distribution of input to a simple distribution, e.g., a Gaussian, GANs use a pair of actors called generator and discriminator that play a min-max game to learn the probability distribution. The components analogous to generator and discriminator of GANs are decoder and encoder in VAEs, respectively. In VAEs, the link between the encoder and decoder is the latent variable distribution, while the link between the discriminator and generator in GANs is the value function, $V(D, G)$, where D and G are the outputs of the discriminator and generator. Unlike the encoder in VAEs, the discriminator in GANs uses the input $X$ and the output from the generator.

The generator generates a sample from the unknown probability distribution of $X$. The discriminator reads two inputs: known input sample, $X$, and the output from the generator to classify it as authentic (i.e., generated from the underlying probability distribution of input) or fake (i.e., generated by the generator). The discriminator attempts to classify the inputs correctly as fake or authentic, that is, reduce the loss function, while the generator tries to produce samples that fool the discriminator, i.e., samples that increase the loss function describing the discriminator's classification. From this min-max game, the generator learns to generate samples that are indistinguishable from the real inputs by the discriminator, at which point the generator is assumed to have learned the probability distribution of input, $X$. The value function (negative of loss function) is the binary cross entropy function for classification into two classes, as shown in equation 6.74:

$$\min_{G}\max_{D} E_{X\sim P(X),Z\sim N(0,1)}\left[\log D(X)+\log\left(1-D(G(Z))\right)\right] \qquad (6.74)$$

During early iterations of the algorithm, the discriminator has little trouble classifying the input $X$ and the output from the generator as authentic or fake. Hence, the term $\log(1-D(G(Z)))$ saturates, with $D(G(Z)) = 0$ with high probability, and the gradients become zero. During these early iterations, it is helpful to modify the function used by the generator for minimization to $\max \log(D(G(Z)))$, as shown in equation 6.75:

$$\min_{G} E_{Z\sim N(0,1)}\left[\log\left(1-D(G(Z))\right)\right]\equiv\max_{G} E_{Z\sim N(0,1)}\left[\log\left(D(G(Z))\right)\right] \quad (6.75)$$

Using a batch size of $M$, the gradient used in stochastic gradient ascent by the discriminator is shown in equation 6.76, while the gradient used in stochastic gradient descent by the generator is shown in equation 6.77:

$$\nabla_\psi \frac{1}{M} \sum_{i=0}^{M-1} \left[ \log\left(D_\psi\left(x = X_i\right)\right) + \log\left(1 - D_\psi\left(x = G_\theta\left(Z_i\right)\right)\right) \right] \qquad (6.76)$$

$$\nabla_\theta \frac{1}{M} \sum_{i=0}^{M-1} \left[ \log\left(1 - D_\psi\left(x = G_\theta\left(Z_i\right)\right)\right) \right] \qquad (6.77)$$

Upon convergence, the generator learns to map the Gaussian distribution of $Z$ to the underlying probability distribution of input $X$ and uses it to generate samples that are indistinguishable from the real inputs by the discriminator. The complete algorithm for training a GAN is sketched in pseudo-code 19.

---

**Algorithm 19**   Generative Adversarial Network Training Algorithm

**Require:** Initial generator network $G(X|\theta, \epsilon)$, initial discriminator network $D(x|\psi)$, mini-batch size $M$, training iterations $N$, discriminator training steps per generator training step, $K$.

1: **for** each n = 0, 1,2,...,N-1 **do**

2:     **for** each k = 0,1,2,...,K-1 **do**

3:         Sample $M$ samples of $Z$ from $N(0,1)$.

4:         Accept $M$ input samples.

5:         Update the discriminator using stochastic gradient ascent with the gradient shown in equation 6.76.

6:    **end for**

7:    Update the generator using stochastic gradient descent
with the gradient shown in equation 6.77. In early
iterations, this can be replaced by using stochastic gradient
ascent on $\mathrm{maxlog}\left(D_{\psi}\left(x=G_{\theta}\left(Z\right)\right)\right)$.

8: **end for**

---

# Bibliography

[1] Hubel, D. H., Wiesel, T. N., *Receptive Fields of Single Neurones in the Cat's Striate Cortex.* Journal of Physiology, 148:574–591, 1959.

[2] Fukushima, K., Neurocognitron: A Self-Organizing Neural Network Model for a Mechanism of Pattern Recognition Unaffected by Shift in Position. Biological Cybernetics, 36:193–202, 1980.

[3] Campbell, J. Y., Shiller, R. J., *The Dividend-Price Ratio and Expectations of Future Dividends and Discount Factors.* The Review of Financial Studies, 1(3):195–228, `www.jstor.org/stable/2961997`, 1988.

[4] White, H., *Economic Prediction Using Neural Networks: The Case of IBM Daily Stock Returns.* Proceedings of the IEEE International Conference on Neural Networks, July 1988.

[5] Barto, A. G., Sutton, R. S., Watkins, C. J. C. H., *Sequential Decision Problems and Neural Networks.* Advances in Neural Information Processing Systems 2, NIPS, 2:686–693, 1989.

[6] Brock, W. A., Lakonishok J., LeBaron B., *Simple Technical Trading Rules and the Stochastic Properties of Stock Returns.* Journal of Finance, 47(5):1731–1764, 1992.

BIBLIOGRAPHY

[7] Williams, R. J., Simple statistical gradient-following algorithms for connectionist reinforcement learning. Machine Learning, 8:229–256, 1992.

[8] Quinlan, R., *Combining Instance-Based and Model-Based Learning*. In Proceedings on the Tenth International Conference of Machine Learning, 236–243, University of Massachusetts, Amherst, Morgan Kaufmann, 1993.

[9] Lakonishok, J., Shleifer, A., Vishny, R. W., *Contrarian Investment, Extrapolation, and Risk*. The Journal of Finance, 49(5):1541–1578, https://doi.org/10.2307/2329262, 1994.

[10] Bradtke, S. J., Barto, A. G., *Linear Least-Squares Algorithms for Temporal Difference Learning*. Machine Learning, 22:33–57, 1996.

[11] Chan, L. K. C., Jegadeesh, N., Lakonishok, J., *Momentum Strategies*. Journal of Finance, 51(5):1681–1713, 1996.

[12] Chenoweth, T., Lee, S. S., Obradovic, Z., *Embedding Technical Analysis into Neural Network Based Trading Systems*. Applied Artificial Intelligence Journal, 10(6):523–541, 1996.

[13] Hochreiter, S., Schmidhuber, J., *Long Short-Term Memory*. Neural Computation, 9(8):1735–1780, DOI: https://doi.org/10.1162/neco.1997.9.8.1735, 1997.

[14] LeCun, Y., Bottou, L., Bengio, Y., Haffner, P., *Gradient-Based Learning Applied to Document Recognition*. Proceedings of the IEEE, November 1998.

[15] Allen, F., Karjalainen, R., *Using Genetic Algorithms to Find Technical Trading Rules.* Journal of Financial Economics, 51:245–271, 1999.

[16] Konda, V., Tsitsiklis, J., *Actor-Critic Algorithms.* Advances in Neural Information Processing Systems, 12, 1999.

[17] Sutton, R. S., McAllester, D., Singh, S., Mansour, Y., *Policy gradient methods for reinforcement learning with function approximation.* Proceedings of the 12th International Conference on Neural Information Processing Systems, 1057–1063, 1999.

[18] Lo, A. W., Mamaysky, H., Wang J., Foundations of Technical Analysis: Computational Algorithms, Statistical Inference, and Empirical Implementation. Journal of Finance, 55(4):1705–1765, 2000.

[19] Kakade, S., *A Natural Policy Gradient.* Proceedings of the 14th International Conference on Neural Information Processing Systems: Natural and Synthetic, NIPS, 01:1531–1538, 2001.

[20] Dreyfus, S., *Richard Bellman on the Birth of Dynamic Programming.* Operations Research, 50(1):48–51, https://doi.org/10.1287/opre.50.1.48.17791, 2002.

[21] Enke, D., Thawornwong, S., *The Use of Data Mining and Neural Networks for Forecasting Stock Market Returns.* Expert Systems with Applications, 29(4):927–940, 2005.

[22]   Leigh, W., Hightower, R., Modani, N., Forecasting
the New York Stock Exchange Composite Index
With Past Price and Interest Rate on Condition of
Volume Spike. Expert Systems with Applications,
28(1):1–8, 2005.

[23]   Savin, G., Weller, P. A., Zvingelis, J., *The Predictive
Power of "Head-and-Shoulders" Price Patterns
in the U.S. Stock Market*. Journal of Financial
Econometrics, 5(2):243–265, 2007.

[24]   Li, S. T., Kuo, S. C., Knowledge discovery in financial
investment for forecasting and trading strategy
through wavelet-based SOM networks. Expert
Systems with Applications, 34(2):935–951, 2008.

[25]   Chavarnakul, T., Enke, D., Intelligent Technical
Analysis Based Equivolume Charting for Stock
Trading Using Neural Networks. Expert Systems
with Applications, 34:1004–1017, 2008.

[26]   Krizhevsky, A., Sutskever, I., Geoffrey, H., *ImageNet
Classification with Deep Convolutional Neural
Networks*. Advances in Neural Information
Processing Systems, NIPS 2012, Vol. 25, 2012.

[27]   Geoffrey, H., *Neural Networks for Machine Learning,
Lecture 6a, Overview of mini-batch gradient descent*,
University of Toronto, 2012.

[28]   Zeiler, M. D., *ADADELTA: An Adaptive Learning
Rate Method*. arXiv, https://doi.org/10.48550/
arxiv.1212.5701, 2012.

[29]   Cho, K., Merri͡enboer, B. V., Gulcehre, C., Bahdanau, D., Bougares, F., Schwenk, H., Bengio, Y., *Learning Phrase Representations Using RNN Encoder-Decoder for Statistical Machine Translation*. arXiv preprint, arXiv: 1406.1078, 2014.

[30]   Goodfellow, I. J., Pouget-Abadie, J., Mirza, M., Xu, B., Warde-Farley, D., Ozair, S., Courville, A., Bengio, Y., *Generative Adversarial Networks*. arXiv: 1406.2661 [stat.ML], https://doi.org/10.48550/arXiv.1406.2661, 2014.

[31]   Kingma, D. P., Ba, J., *Adam: A Method for Stochastic Optimization*. arXiv, https://doi.org/10.48550/arxiv.1412.6980, 2014.

[32]   Kingma, D. P., Welling, M., *Auto-encoding Variational Bayes*. International Conference on Learning Representations 2014, arXiv: 1312.6114 [stat.ML], https://doi.org/10.48550/arXiv.1312.6114, 2014.

[33]   Silver, D., Lever, G., Heess, N., Degris, T., Wierstra, D., Riedmiller, M., *Deterministic Policy Gradient Algorithms*. Proceedings of the 31st International Conference on Machine Learning, DOI: 10.5555/3044805.3044850, 32:387–395, 2014.

[34]   Fama, E. F., French, K. R., *A Five-Factor Asset Pricing Model*. Journal of Financial Economics, 116(1):1–22, DOI: https://doi.org/10.1016/j.jfineco.2014.10.010, 2015.

[35]   Schaul, T., Quan, J., Antonoglou, I., Silver, D., *Prioritized Experience Replay*. http://arxiv.org/abs/1511.05952, arXiv: 1511.05952, 2015.

[36]   Schulman, J., Levine, S., Moritz, P., Jordan, M.,
       Abbeel, P., *Trust Region Policy Optimization*.
       Proceedings of the 32nd International Conference
       on Machine Learning, 37, ICML 15:1889–1897, DOI:
       10.5555/3045118.3045319, 2015.

[37]   Ahlawat, S., Empirical Evaluation of Price-Based
       Technical Patterns Using Probabilistic Neural
       Networks. Algorithmic Finance, 5(3-4):49–68, 2016.

[38]   Dozat, T., *Incorporating Nesterov Momentum into
       Adam*. Workshop track – ICLR, 2016.

[39]   Hasselt, H. V., Guez, A., Silver, D., *Deep
       reinforcement learning with double Q-learning*.
       AAAI'16: Proceedings of the 30th AAAI Conference
       on Artificial Intelligence, 2094–2100, 2016.

[40]   Lillicrap, T. P., Hunt, J. J., Pritzel, A., Heess, N., Erez,
       T., Tassa, Y., Silver, D., Wierstra, D., *Continuous
       Control with Deep Reinforcement Learning*.
       Computing Research Repository, http://
       dblp.uni-trier.de/db/conf/iclr/iclr2016.
       html#LillicrapHPHETS15, 2016.

[41]   Wang, Z., Schaul, T., Hessel, M., Hasselt,
       H. V., Lanctot, M., De Freitas, N., *Dueling
       Network Architectures for Deep Reinforcement
       Learning*. ICML 16: Proceedings of the 33rd
       International Conference on Machine Learning,
       48:1995–2003, 2016.

[42]   Fortunato, M., Azar, M. G., Piot, B., Menick, J.,
       Osband, I., Graves, A., Mnih, V., Munos, R., Hassabis,
       D., Pietquin, O., Blundell, C., Legg, S., *Noisy
       Networks for Exploration*. arXiv: 1706.10295, 2017.

[43]  Schulman, J., Wolski, F., Dhariwal, P., Radford, R., Klimov, O., *Proximal Policy Optimization Algorithms*. ArXiv, abs/1707.06347, 2017.

[44]  Barth-Maron, G., Hoffman, M. W., Budden, D., Dabney, W., Horgan, D., Dhruva, Muldal, A., Heess, N., Lillicrap, T., *Distributional Policy Gradients*. International Conference on Learning Representations, `https://openreview.net/forum?id=SyZipzbCb`, 2018.

[45]  Fujimoto, S., Van Hoof, H., Meger, D., *Addressing function approximation error in actor-critic methods*. arXiv preprint, arXiv: 1802.09477, 2018.

[46]  Haarnoja, T., Zhou, A., Abbeel, P., Levine, S., *Soft Actor-Critic: Off-Policy Maximum Entropy Deep Reinforcement Learning with a Stochastic Actor*. Proceedings of the 35th International Conference on Machine Learning, 80:1861–1870, `http://proceedings.mlr.press/v80/haarnoja18b/haarnoja18b.pdf`, 2018.

[47]  Roy, P., Ghosh, S., Bhattacharya, S., Pal, U., *Effects of Degradations on Deep Neural Network Architectures*. arXiv preprint. arXiv: 1807.10108, 2018.

[48]  Koklu, M., Ozkan, I.A., *Multiclass Classification of Dry Beans Using Computer Vision and Machine Learning Techniques*. Computers and Electronics in Agriculture, Vol. 174, 105507, DOI: `https://doi.org/10.1016/j.compag.2020.105507`, 2020.

[49] Kaggle, *Credit Card Fraud Dataset*. Available at `www.kaggle.com/datasets/mlg-ulb/creditcardfraud`. Accessed June 1, 2022.

[50] UCI, *Dry Bean Dataset*. Available at `https://archive.ics.uci.edu/ml/datasets/Dry+Bean+Dataset`. Accessed June 1, 2022.

[51] UCI, *Auto MPG Dataset*. Available at `https://archive.ics.uci.edu/ml/datasets/Auto+MPG`. Accessed June 1, 2022.

# Index

## A

Accuracy, 44, 46, 47, 49, 173
Activation gate, 187
Actor-critic algorithms, 326–329
    agents, 328
    correlation coefficient, 331
    PNL, 333, 334
    policy function, 328
    policy optimization, 328
    relative weights, 332
    reward, 332
    Sharpe ratio, 333
    softmax layer, 332
    Stochastic gradient, 329
    TD target, 328
    testing, 333
    trading strategy
        buy-and-hold strategy,
            334–336, 338, 339,
            341, 343–345
        daily log, 330
        portfolio holding period, 330
        stocks, 330
    training data, 332
    transaction costs, 332
    value function, 328
    variance ratio, 330, 331

American barrier option
    actions, 287
    asset price, 287
    *vs.* European option, 285
    nodes, 287
    option price
        contour plot, 294, 296
        publicly traded stock, 287
        surface, 294, 295
    SARSA algorithm
        computed price, 294
        optimal policy, 286
        option price, 288–290,
            292, 293
    state-action value function, 287
    state space, 287
    state value function, 287
    stock price/strike price, 287
Artificial intelligence algorithms, 1
Automated trading strategies, 3
Autoregressive model, 208
AveragePooling1D, 23

## B

Backpropagation through time
    (BPTT), 177
BatchNormalization, 23

Printed in the United States
by Baker & Taylor Publisher Services